ARE YOU A MACHINE?

THE BRAIN, THE MIND, AND WHAT IT MEANS TO BE HUMAN

ELIEZER J. STERNBERG

ILLUSTRATIONS BY SHANNON BALKE

Humanity Books

an imprint of Prometheus Books
59 John Glenn Drive, Amherst, New York 14228-2197

Published 2007 by Humanity Books, an imprint of Prometheus Books

Inquiries should be addressed to
Humanity Books
59 John Glenn Drive
Amherst, New York 14228-2197
VOICE: 716-691-0133, ext. 207
FAX: 716-564-2711

11 10 09 08 07   5 4 3 2 1

Library of Congress Cataloging-in-Publication Data

Sternberg, Eliezer J.
    Are you a machine? : the brain, the mind, and what it means to be human /
Eliezer J. Sternberg.
        p. cm.
    Includes bibliographical references.
    ISBN 978-1-59102-483-5 (alk. paper)
    1. Philosophical anthropology. 2. Human beings. I. Title.

BD450.S7657 2007
128—dc22

2006032659

Printed in the United States of America on acid-free paper

For my parents,
to whom I owe everything.

CONTENTS

8 **Contents**

FOREWORD

For many, if not most, the idea that we are machines is thought provoking, yet hard to believe. It's hard to believe because most of us think there is more to our being human than the merely mechanical. We are special.

MINDS AND MACHINES

We think we are special in part because we have an image of ourselves and an image of machines that refuse to line up. Some might say we are overly impressed by ourselves and underwhelmed by machines. But putting aside for the moment our attitudes about ourselves, we tend to "see" machines as mechanical devices constructed out of "hard stuff"—inorganic material—rather than organic, squishy stuff. Organic and squishy is not the same as hard and material. So (obviously) we are not machines.

But it's not just the hard stuff. What many believe sets us apart from machines is the fact that we are able to engage in kinds of activities that are permanently closed off to machines: we have a mental life. We are conscious of the world inside and around us. Machines are not.

There was a time, not so long ago, however, in the history of the biological sciences when life itself was believed to be a mystery and could only be understood by the presence of something that was itself quite mysterious: a *vitality* or *life-force*. But this *vitality* is now explained in quite mechanical terms. Life is understood as being made up of a number of underlying organic processes, metabolic and reproductive processes, to name a couple, that do not require any "added" ingredient, some "mysterious" something or other, a *vitalism*, to bring these processes to life, to make them *live*.

Still, to many, the human mind presents a radically different, far deeper, if not impossible, challenge. It is one thing to explain metabolism in mechanical terms; quite another to explain human consciousness in any way that might be regarded as somehow "equivalent." Despite progress in the life sciences, many believe an explanation of human consciousness will remain beyond our grasp.

COMPUTERS

One of several developments in the last half century, however, has caused some to have second thoughts and renewed hope that a naturalist account of human consciousness may, after all, be within reach. I am thinking of computers and the dramatic developments in the areas of information processing, artificial intelligence, and computer programming.

Computers can perform tasks, for example, draw inferences, move from premises to conclusions, and recognize patterns, thoughtlike tasks that we heretofore regarded as unique to us. Indeed, the design and modeling of computers have become so sophisticated that more and more complex mental activities are programmable as computer operations.

If we can wean ourselves of our attachment to the belief that we, unlike machines, are made out of organic, "living" stuff and see ourselves rather as creatures with certain capacities, the difference between us and machines may be seen to rest on our ability to perform tasks machines cannot. The difference between machines and us would lie not in the "stuff" out of which we are made, but in the tasks we are able to perform. By the same token, machines can then be seen as more or less like us in light of how well they are able to do what we do. Any suitable component that made it possible for a machine to perform a given task would make it functionally equivalent.

An artificial heart *is* a heart if it does a heart's job. Whether it be made out of plastic or organic matter, a heart *is* as a heart *does*. It's a heart if it can do the work of a heart. A similar point might be made about machines. A machine is thinking if it can perform the job that thinkers do. It can be made out of any stuff whatsoever so long as its construction enables it to play the role appropriate to someone who thinks.

These and similar reflections coupled with the idea that human beings are information-processing systems helped to create the field of Artificial Intelligence or "AI," whose research program is essentially one of getting computing machines to do what only human thinking and intelligence have heretofore been able to do. AI provokes us to wonder to what extent (A) computers are like minds and (B) minds are like computers.

Although both A and B raise their own set of questions and concerns, the two issues are connected. By exhibiting capacities that have conventionally been understood as belonging solely to the sphere of the human mind, computers have not only encouraged the belief that human consciousness can be explained, but in their very operation serve to confirm that a mechanistic understanding is possible, since computers have, as Daniel Dennett has noted, "nothing up their sleeves," nothing all that mysterious to hide.

In probing A we might wish to know more specifically what mental activities computers can perform. In thinking about what computers *can* do, we discover that often we are asking two different sorts of question: (1) what can computers do *in practice* and (2) what might they be able to do *in principle?*

Both sorts of question arise in the chapters that follow. But even if we agree that computers are capable of performing all kinds of complicated mental tasks, we would still want to know if computers performed these tasks in the same way that human beings perform them. Even more important, we might wonder if computers performed quite sophisticated psychological tasks in the same way human minds performed them, would that show— beyond a doubt—that computers were conscious.

In switching from the A-side to the B-side of this issue, that is, in thinking about the extent to which human minds are like computers, we might well wonder, to adapt a phrase from William Lycan: "Can human psychological capacities be entirely captured by a third-person, hardware-realizable design of *some* sort that could in principle be built in a laboratory?"

To date, computers have been moderately successful in what tasks they can perform. We might suspect, however, if machines can "draw inferences," what more can they do? Couldn't they be programmed to do more and more until they were seen to think like us and, once they were thinking like us, couldn't machines be said to be conscious? If "consciousness" best describes the missing ingredient between them and us, between machines and human beings, perhaps we are not so special.

CONSCIOUSNESS

Elie Sternberg has ostensibly written a book about machines, but in fact his book gives equal, if not greater, weight to a topic that some believe to be the "last great mystery" of modern science. That "great mystery" is the mystery of human consciousness. Whether it is the last of the great mysteries is perhaps open to question—surely there will be others—but there is widespread agreement among scientists and philosophers that human consciousness is a great mystery.

Despite this agreement there is surprisingly little agreement on what that mystery consists in. To this disagreement one might respond, "Yes, indeed, that is why it is such a mystery," but most mysteries have a form and can be couched in a set of questions that everyone agrees are the fundamental questions to be asked and answered. As Sternberg's book makes plain, the study of human consciousness is still in its infancy. As such it is far from settled what the basic questions to be asked in this field are.

PHILOSOPHY AND NEUROSCIENCE

Largely because the study of human consciousness is "up for grabs," as it were, philosophical inquiry has played no small role in the early stages of its development. Philosophers are able to contribute to what on its face might otherwise appear to be a wholly scientific question by helping to frame the questions.

Philosophy and science are similar sorts of enterprise in part because neither is limited by a particular subject matter in the way; for example, economics is restricted to a circumscribed area of human endeavor. In this respect both philosophy and science are open-ended in their scope and interest. But they differ from one another in their approach, in particular in the way each asks questions and goes about answering them.

Science works best when there are phenomena or a phenomenon that it can examine in a systematic way. Philosophy tends to investigate questions where there is yet to be a systematic method for finding an answer. Since the study of human consciousness is concerned with questions where there is no agreed-upon method for going about getting and giving answers, the neuroscience is ripe for philosophy and somewhat bereft without it.

There is a sense that philosophy stands in relation to science as a form of pre-science, a kind of preliminary work before the science gets wholeheartedly underway. Like an overture that precedes the opera, philosophy begins to set the stage, lay the groundwork, and pick out the themes to be explored in greater depth in the science to come.

So, too, philosophy is often practiced as a form of conceptual analysis. At the beginnings of a new science, conceptual issues abound, and that is especially true in the study of human consciousness. Philosophy is inclined to step back and examine an issue because it pays to do so. Unless one steps back and looks more closely at the ways one is conceiving a problem, one may be misled by one's way of seeing that problem rather than by one's inability to solve the problem. Indeed, if one is conceiving of a problem in a misleading or wrongheaded way, the last thing one should want is a solution. What one should want is a new way of looking at that problem, of conceiving it in some *other* way rather than in *this* or *that* way. Viewed in this light, the philosophy and science of human consciousness were made for each other.

Sometimes, too, what is required is a better understanding of the key concepts that are driving our investigation of a phenomenon. Here philosophy is an aid to the neuroscience as well. Just as a question such as "Is a minimum wage fair?" will benefit from a better understanding of what we mean by "justice," so, too, the temptation to identify consciousness as, among other things, "private" and "subjective" is likely to benefit from a better grasp of what we mean by what is "private" and what is "subjective."

The significance of the results of experiments in the early stages of the development of a new science, in this case in the early stages of a new science of consciousness, is debatable. Details of a new science cry out for interpretation, and the scientists can and will interpret their own results and often do so with great enthusiasm. But in the early stages the interpretations of results provided by the investigators themselves are usually widely speculative. In this regard the investigators need all the help they can get, and here, once again, philosophy can play a useful role in seeing how the brain science fits in with human experience, with introspective talk of mental activity, with, for example, human thinking and feeling.

AN INVITATION TO THINK

The aim of this book, however, is less to cover ground than to introduce readers to what's puzzling about human consciousness and to do it in such a way that encourages readers themselves to be puzzled. In this respect the book is an attempt to enlist recruits to join an exhilarating conversation about the mind and the brain.

In consequence, Sternberg's fifteen chapters are less an introduction than an invitation. Invitations "beckon" rather than "point the way." Sternberg's book does provide instruction, but, more important, it summons the reader to think about the problem of consciousness, about its nature and causes, and about whether and how it might ultimately be understood.

Thus the book is less a survey of past results or an overview of what has been accomplished to date than a "call" to think. It does not attempt to be comprehensive or exhaustive. Sternberg is aware that facts matter, but the book is not a litany of scientific facts. Its primary aim is to provoke readers to think imaginatively and creatively about the relation of mind and brain, to come up with new ways of thinking, in particular, thinking of their own, to see if the mysteries associated with human consciousness can be resolved. To jump-start this process, Sternberg relies less on clumps of data than on analogies, stories, and thought experiments designed to capture readers' imaginations, pull them in, and bring them to a point where they are open at least to revising their unreflective views of consciousness and replacing those views with new visions of the mind.

The thought experiments that Sternberg deploys in this project are intended in a similar spirit to test our intuitions as well as reshape them. The thought experiments in the book, such as the "Chinese Room" argument or the "Zombie" thought experiment, function a little like cloud-chambers enabling readers to "track" their intuitive understandings of certain puzzling features of the human mind and to follow their trajectory to a place where readers may (perhaps) be ready to *rethink* what they had intuitively accepted up to that point.

The book, then, does not tell the reader what to think; it encourages the reader to think. In both its structure and spirit it represents what is best in contemporary philosophy. As J. R. Lucas has said:

> Philosophy has to be self-thought, if it is to be thought at all. It is an activity rather than a set of positions. You need to think out the problems and solutions for yourself, and although another person's philosophizing may help you in your own, you cannot accept their conclusions, or even understand their arguments, until you have already argued a lot with yourself.

TAKING A POSITION

At the end of the book Sternberg gives his own view, so the book has an ending of sorts. But that ending still leaves room for the reader to form his or her own view. Sternberg's view does not occupy all the remaining theoretical territory. The view is presented clearly and simply, but in a way that allows readers to raise questions and, as I have said, form different opinions.

As I have been suggesting, the pursuit of an explanation of consciousness is challenging in part because there is less than full agreement on what is to be explained (the *explanans*) and what would be required to explain it (the *explanandum*).

What is unique and special about the thinking in the book is not just its complexity. It also serves as a reminder of what a theory of consciousness would need to explain. The thinking that takes place between the covers of the book is an example of an aspect of human consciousness that a theory, were such a theory to exist, should be able to capture.

The book is therefore about the promise of a theory of consciousness as well as the obstacles that might be seen to stand in the way of its development. Are the obstacles real, real enough? And might they be overcome?

Is a theory within our grasp or do we lack the know-how and the imagination to penetrate the mystery of consciousness? Perhaps we are just not smart enough. Or perhaps creatures like us who are conscious ourselves cannot understand the nature of our own consciousness.

If you do not already have a view on this, by the end of Sternberg's book you are likely to have an answer. It may not be the correct answer. But if it is not correct it is not likely you will know that for some time to come. What will be true is that the answer you give will be *your* answer. That may be little consolation if it turns out that it is incorrect. But if there is any consolation to be had here, one way, perhaps the only way, to coming to understand consciousness is by asking and answering questions, by taking positions, seeing how well the positions hold up, and abandoning them when they give out. The process of question and answer is surely the best and most reliable route to solving the problem of human consciousness.

The strength of Sternberg's book rests in its invitation to its readers to begin this process, in its invitation to look for and find solutions to a hard, if not intractable, problem.

Andreas Teuber
Department Chair
Professor of Philosophy
Department of Philosophy
Brandeis University
Waltham, Massachusetts

ACKNOWLEDGMENTS

I wanted to take this opportunity to express my gratitude to a number of people for helping me accomplish what has been a dream of mine for some time.

The first, without question, is my teacher, Carol Palm. It was during my junior year at Williamsville North High School in Buffalo, New York, that Ms. Palm, my English teacher, assigned the class a final writing project. I decided to write a twenty-some-page essay called "Are You a Machine?"

I remember bringing it to Ms. Palm for comments. She looked through it briefly and told me that I should cut out the first five pages. After sulking and muttering under my breath for a while, I found myself sitting in front of my computer at home highlighting pages one through five and using a trembling finger to press the delete button. The following summer, Prometheus Books sent me a letter expressing interest in looking at an expanded version of this essay, which I had sent them.

Much of my senior year of high school was spent at a desk in Ms. Palm's room, delving into the literature of consciousness. Each day, students would walk into the room and find the two of us engrossed in a conversation about the nature of mind, free will, zombies, or, possibly, the neurobiological correlates of the self. It was through these discussions that I solidified my understanding of the ideas in this book. Ms. Palm has had an influence on me more profound than even she realizes. I was privileged to be her student and continue to be so as her friend.

I would like to thank the Kadimah School of Buffalo, and my English teacher there, Lois Anneler, for showing me that I love to write, and Williamsville North High School, for giving me the opportunity to learn from three other wonderful teachers of English: Melanie Metz, Regina Forni, and Mary Richert. My thanks also go to the Williamsville North for not noticing the many, many pages of text that I printed using the library printers.

My friend and fellow graduate Shannon Balke is responsible for the illustrations. Shannon is a marvelously talented artist and a brilliant, hardworking student. She did the drawings while taking a very difficult course schedule and playing varsity sports at the University of Pittsburgh. I am greatly indebted to her for the enormous amount of time that she put into this project.

I made the wise choice to attend Brandeis University, where I am currently studying. Brandeis has provided me with a wonderful environment to learn and grow. It was here that I met Professor Andreas Teuber, who has written the foreword to this book.

Professor Teuber, currently the chair of the philosophy department at Brandeis, has helped me take my study of philosophy and neuroscience in new directions. Through our hours of stimulating interchange, he has shown me the depths that can be reached in philosophical interaction. What's more, he took upon himself the task of reading over my manuscript several times before returning it to me with comments. It is humbling to think that he did so much for me without yet having had me as a student.

The Brandeis philosophy department is very lucky to have Julie Seeger as its academic administrator. She helped me by tracking down all the sources for photographs (a tremendously complicated task) and by tolerating my constant visits to her office.

I am also grateful for the help of University of Oregon professor Robert Sylwester, who gave me so much of his time without ever having met me. He and I had many exchanges by e-mail in which he would pass on sources of information, point me to interesting books and articles, advise me in my pursuit of the neurosciences, and send commentary on my chapters. He has written a great many letters on my behalf and has been an unwavering source of support.

A number of others helped me with their comments. I owe gratitude to Brandeis professors of philosophy Eli Hirsch and Jerry Samet, for their suggestions, and to neuroscience professor Robert Sekuler for making sure I had my science right. I also want to thank my friends Emily Zevon, Nechama Horwitz, Josh Balderman, Jason Rothwax, Avi Bieler, Charlie Gandelman, Jessica Kent, Esti Schloss, Sara Smith, and Hanna Cohen for looking at the manuscript and giving their thoughts and, more important, encouraging me to pursue my interests.

Most of all I want to thank my family. My greatest blessing is to have been the son of Ernest and Zohara Sternberg and the brother of Daniel, Benjamin ("Benny," star of chapter 4), and Rebecca. If it turns out in the end that we are machines, these machines will need no upgrades.

Elie Sternberg
December 2006

INTRODUCTION

I know a world that no one else knows. A key is needed to get inside it and, insofar as I can tell, I'm the only one who holds that key. I'm talking about the world of my inner thoughts, and the key I use to enter it is thought itself. I seem to be the only one who can wander through this world so rich with ideas, opinions, memories, and experiences.

Two years ago, I began to grow curious about how this mental world works and how it is possible. When I sought explanations, I was naturally led to biological accounts of brain functions. In fact, textbooks on the brain hold that a better understanding of neuronal interactions will eventually reveal the secrets of the mind. Somehow, science will reveal how the matter in my skull will give me access to a boundless mental life.

If so, thoughts must be products of the brain just as bile is of the liver and saliva of glands in the mouth. I received this explanation with discomfort. If my thoughts are reducible to physical processes in my brain, what would distinguish me from a robot built to process information and behave as I do? If not in my mind, where does my humanity lie?

As I was considering this question I happened upon an article by philosopher John Searle. Titled "Is the Brain's Mind a Computer Program?" it was published in *Scientific American*. In his article Searle presented arguments for why programmed computers cannot make sense of the world as the human mind does.

Through his article and the readings that followed it, I was introduced to the enormous body of literature on the question of human consciousness. It turns out that philosophers, even those aware of modern advances in the sciences, do not at all agree that knowledge of the brain will allow us to predict thoughts as if they were results of a mechanical process. In fact, philosophers hold a variety of fascinating and conflicting views. Their writings show that, in light of recent developments in neuroscience and artificial intelligence, ancient questions on the relation between mind and body have taken on a new salience. Today, ever more lines of philosophical investigation, biological research, and computer modeling are converging on the great mystery of human consciousness.

Many researchers believe that a complete knowledge of how the brain works is not far from our reach and that we will soon be able to access the contents of human minds. Others see the future as bringing the development of conscious machines that will challenge our conception of ourselves as unique creatures. Still others argue that the mental world has capacities for free will, reasoning, and imagination that technological advances will be unable to invade. Much rests on the answer to the mystery.

What further drew me to this debate is that it is rich with examples, thought experiments, and just juicy ideas. The Turing Test, the Chinese Room, animal consciousness, zombies, ghosts in the machine, alternate worlds, conscious machines—they make the field come alive, even to those of us untrained in philosophy. It is through such ideas that leading writers debate whether advancements in science and technology will show that we human beings are nothing more than machines: highly sophisticated machines to be sure, but machines nonetheless.

I became engrossed in this debate in search of answers about myself, but it turned out that the debate was in itself fascinating. The fifteen chapters that follow will seek to convey what is so special about the controversy. It is only in the last chapter that I try to fashion my own position. Throughout, my intent is to reveal that the debate matters not only for abstract philosophical exchange, but also for our conception of ourselves.

1

IN THE SCIENTIST'S LAIR

Suppose that a scientist has mastered the workings of your brain. After years of study, using powerful scanning technology, he has examined every cell, every neural connection, and every chemical interaction that exists within your skull. He knows precisely how all these different parts work together. On a giant computer screen, he displays a minutely detailed and extremely accurate digital model of your gray matter. The scientist, however, claims to know more than just the structure of your brain. He declares that he is an expert on how your brain generates thought, how you make decisions, and how you behave, even though he has never seen you outside his laboratory. The scientist says that he understands everything about your mind—your memories, doubts, fears, hatreds, cares, sorrows, and beliefs—all from studying the structure of your brain.

Do you think that what he claims is possible? The scientist maintains that the brain is an organ like any other. It is a mechanical apparatus just like the heart or the liver or the stomach. "We can understand everything about how *those* organs function," he says, "so why should the brain be any different?"

If the scientist knows everything there is to know about your brain, would he know what you are thinking at any moment? Would he be able to predict your actions? Would he truly know everything about your mind? Would he know how it feels to have your perspective? Would he know *what it is like* to be you? If he truly could understand all of this just from studying the mechanisms in your brain, would that make you a machine?

Before we begin to consider this question, we must define what a machine is. The word *machine* is used in a number of ways. Levers and pulleys are called machines, as are cars and motorcycles. Computers and robots are machines, as well as popcorn makers and toaster ovens. The human heart and liver are machines too. A machine can be made of any material. We will consider a machine to be *a system of interacting physical parts that operates according to a set of formal rules to accomplish work*. So, a lever is a machine because it is a system of parts (a bar and a fulcrum) that interact according to a set of rules: when one side of the bar goes down, the other side goes up. The work the lever performs is lifting one side of the bar (no wonder they call it a "simple" machine), which usually has an object on it. Keep this definition of a machine in mind when considering the questions to come.

Now imagine that the scientist uses his digital model on the computer screen to physically re-create your mind. He first develops silicon chips to duplicate the functions of the neurons in your brain. Meticulously, he reconstructs your neural network, with all its signal pathways intact, using specially manufactured, intricate circuitry. Next, the scientist programs the various parts to operate in precisely the same way that your brain does. He also programs the system with all the knowledge and memories that you possess. If the scientist builds it with structure and operation equivalent to yours, would the machine be you?

The machine processes information in the same way that your brain does and has your physical structure, but would it have your thoughts? Your emotions? Would the *machine* experience the world the way you do? If so, it would be conscious, but how would you be able to tell? Is there any way to test for consciousness? A robot might tell you all about its inner feelings and its childhood and, when you ask it follow-up questions, it might give satisfactory answers. It might cry when you strike it, grimace when you insult it, and grin when you compliment it. Would that make it conscious?

If the scientist *could* build a machine that is truly conscious, that would mean that consciousness is the product of completely mechanical operations. It might mean that we, too, are machines.

Human consciousness is probably the greatest remaining mystery. As much as we think we know about the brain, nobody has ever figured out how it provides us with consciousness. Most people agree that we could not be conscious without a brain, but exactly how does it *make us* conscious? How does each of us have an identity? What is the self? Where does free will come from? What exists within us that decides whether to have chicken or fish for dinner, feels pain and love, and views the world with one unified perspective?

Some say that the answer to all of this is completely mechanical—that the human mind is merely a computerlike system of signals and responses and is not that different from other natural processes such as digestion or photosynthesis. If this is true, then the scientist should be able to learn everything about your consciousness by simply studying the structure of the brain that generates it. This could mean that we are machines.

Others believe that consciousness is something distinct from the physical world—that no matter how much the scientist may know about your brain, he cannot know what you are thinking. The mind is something fundamentally different from physical processes, not merely a mechanical function. Some think that human consciousness is beyond the reach of scientific study altogether. After all, despite many years of study, nobody yet understands how consciousness is created.

In recent years, because of great leaps in technology and scientific discovery, the age-old study of consciousness has taken new turns. Computers are becoming more and more powerful. Research associated with various neurological disorders is hinting at the operations of some of the systems of the brain. Robots in artificial intelligence labs are accomplishing impressive new feats. Innovative philosophical theories are being proposed. The study of consciousness has become the junction between philosophy, psychology, cognitive science, neuroscience, computer science, and engineering. Experts from these fields now grapple with our questions of whether a conscious machine could be created and what the implications of such an accomplishment would be for the future of humanity.

Can the scientist understand everything about your mind by studying the structure of your brain? Can he build a conscious machine? What is consciousness, anyway? In the chapters that follow, we will examine the ideas developed by a number of major players in the controversy over human consciousness as we try to answer the question: *Are you a machine?*

FURTHER READINGS

This book does not focus on consciousness alone, but on how consciousness relates to the question of whether we are machines. If you are interested in exploring an in-depth philosophical account of consciousness, look for *The Rediscovery of the Mind* by philosopher John Searle. For a psychological perspective, try *In Defense of Human Consciousness* by Joseph Rychlak. To learn about the biology of consciousness, read *Consciousness* by J. Allan Hobson and *Consciousness: A User's Guide* by Adam Zeman. A good summary of many views on consciousness, meant as an introductory college textbook, can be found in *Consciousness: An Introduction* by Susan Blackmore. For the publication information about books and articles mentioned here—or other Further Readings sections—see the bibliography at the end of the book. To explore the question of whether we are machines, turn the page.

2

THE MYSTERIOUS POWER

It is morning in London, and, in his home at 221B Baker Street, Sherlock Holmes wakes up. Minutes ago, he was sleeping, unaware of his surroundings. Now that he is awake, his awareness returns. In a way, he is *conscious*.

Today, Watson has come to visit Holmes, who appears to be deep in thought as he sits at the breakfast table. Watson looks at him. "Good morning, Holmes."

Distracted by his thoughts, Holmes is not paying attention to Watson. You might say that he is not *conscious* of the fact that Watson is there, speaking to him. Watson taps Holmes on the shoulder and repeats himself. Now Holmes has heard him. "Oh, hello there, dear Watson. Good morning. What brings you to my door?"

"A problem, I'm sorry to say."

"What sort of problem?"

"Well, for years I have held on to a very special antique, given to me by my father. It is a nugget of solid gold in excellent condition. I'm afraid I lost it yesterday afternoon."

"How did you do that?"

"It was rather peculiar, actually. I was on my way to the bank to put the nugget in my safe deposit box when, as I was walking past the park, I decided to take a short stroll. There was a bench nearby, so I sat on it. I took the gold nugget from its brown leather case to admire it. The gleam of the gold in the sunlight was so dazzling that I barely heard my phone ring. Startled, I put the nugget in its case and set it on the bench. I took my phone out of my pocket and answered it. The reception was poor so I walked about ten feet from the bench to talk. Before I did that, though, I checked to make sure that nobody was in the park. I was on the phone for only a minute before I returned to my seat and found that the nugget had vanished."

"Hmm . . . interesting," says Holmes. "You are sure that there was nobody else around who could have stolen it?"

"I am certain. The park was empty. I would have seen anybody approaching or running away. What do you make of it, Holmes?"

"Watson, do you recall the area in which you sat?"

"I do, Holmes. I can visualize it now."

"Take this paper and draw what you see in your mind."

Watson pulls a pen from his pocket and begins to draw. He sketches the bench standing by an oak tree in a grassy field. When finished, he hands the drawing to Holmes.

"My dear Watson, I know the location of your lost gold nugget."

"Where?"

Holmes draws an "X" on the paper. "It is here, between the bench and the tree, about three inches under the soil."

"How could you know that?"

"Elementary, my dear Watson. It was not a person who caused your precious nugget to vanish, but a squirrel. Mistaking the round, brown leather-encased antique for a nut, the little creature snatched the nugget from the bench and buried it several feet from the tree. Here is a shovel, my friend. Go recover your father's heirloom."

A tear at the corner of his eye, Watson thanks Holmes graciously and leaves.

Consciousness is a notion central to the question of whether we are machines. It is, however, a difficult concept to define. In the above scenario, Holmes and Watson demonstrate both what we mean by consciousness and what we don't mean.

The word *consciousness* is used in many ways in everyday speech. When Holmes wakes up, we call him conscious, meaning awake. When he isn't paying attention to Watson, we say that he isn't "conscious" of the fact that Watson is there, speaking to him. These are two common uses of the word. In addition, if Holmes were to enter a boxing match, he could be knocked *unconscious*. If he sustains an injury, he could be unconscious for a long time. He could fall into a coma. After receiving treatment, he would wake up from the coma and, once again, be conscious. This is a third use of the word. On any normal day, Sherlock Holmes will be awake, paying attention and not in a coma, but this is not the only reason for saying that Holmes is conscious. In the course of their exchange, Holmes and Watson demonstrate many other reasons.

During their conversation, they display a major feature of consciousness: the capacity for language and understanding. Holmes and Watson are not just mouthing and receiving phrases, but are attaching *meaning* to the words they say and hear. At one point in the discussion, Watson explains that as he sat on the bench, he admired the gleam of his gold nugget in the sun. Watson didn't just detect the shining color; he appreciated and was pleased by it. He had the conscious experience of seeing bright gold. This is another aspect of consciousness: the ability to have private, inner experiences like that one—to have a *perspective* and mental existence. Associated with this is the power to *imagine*, just as Watson imagines the scene of the bench by the tree.

When Holmes uses Watson's story to figure out what happened to the nugget, he demonstrates the power of human *reasoning*. He is able to analyze the evidence presented to him and form an explanation of what happened. This brings us to two other features of consciousness: the self and free will. The *self* is what does, or appears to be doing, the thinking, what makes decisions and has *intentions*. It is our identity, what we each refer to when we say "I." *Free will* is the ability of the self to control thought and the body. The human mind (note that the word *mind* refers to consciousness, not the brain) allows us to wield free control (to a point, many will say) over what we think and what we do.

Finally, the capacity for *emotion* is also an important part of consciousness. When Watson is told where his father's nugget is, he feels more than just the tears in his eyes. He has interior, private experiences of relief, joy, and possibly affection for the man who just helped him. So, between the two of them, Holmes and Watson demonstrate the powers of language, understanding, experience, perspective, imagination, thinking, the self, intention, free will, and emotion all in one short scene. Together, these abilities form the mysterious power of consciousness.

FURTHER READINGS

For a discussion of various aspects of consciousness, look for the essay "Consciousness" by philosopher John Searle. It is widely available online. An interesting book to check out is *The Mysterious Flame* by Colin McGinn. Also relevant is *In the Theater of Consciousness: The Workspace of the Mind* by Bernard Baars.

3
THE GHOST
IN THE MACHINE

Basketball tryouts are coming up and I am trying to decide whether or not to join the team. Being on a sports team is time-consuming, but I enjoy playing. I can see myself in the action now. The ball is passed in to the point guard. Beads of sweat on his forehead, he crouches with the ball and sweeps the court with his eyes like a cat searching for its prey. Suddenly, he sticks out his left foot in a fake and throws his weight to the right. With two dribbles, he is past his defender and heading toward the hoop. My defender drifts away from me, trying to stop the man who is dribbling. I call for the ball. The point guard sees me and fires the ball behind his back, sending it straight toward me. I catch it. The ball feels solid in my hands. With three quick strides, I drive along the baseline and neatly lay the ball off the backboard to score.

No, this scene is wrong. That is not what would happen. I see the ball passed in to the point guard and the agility with which he passes his defender. He dribbles toward the basket. A defender has drifted in front of him. He sees a teammate and fires the ball to him behind his back. His teammate then neatly lays the ball off the backboard. I am in this scene, but not on the court. I am watching others play from the bench, where I have been since the game began. This scene seems the more likely of the two. Maybe joining the team isn't such a great idea.

The conscious mind is extraordinary. It is completely different from all other natural phenomena because none of the rules that govern the physical world seem to exist there. There are no laws of physics and no limits of possibility. In the physical world, I might only be able to jump a foot high, but in my mind, I can jump well over the rim, do several flips in the air, and dunk the basketball with my feet. In my mind, I am in control. I have what seems to be an endless power to change what goes on in my imagination. I can instantly change scenes in my imaginary world, as from an enjoyable to a miserable experience of basketball. Imagination has no boundaries.

Nobody can ever know what occurs in my mind because my consciousness is private. Not even science has access to it. A scientist will not know that I am thinking about basketball unless I tell him. Science explains how things in the physical world function, such as the way the digestive system works or how water turns to ice. It requires observable data to make calculations and hypotheses. Modern science is unable to explain consciousness in such terms. Consciousness is truly a mysterious power.

Philosopher David Chalmers writes: "Consciousness poses the most baffling problems in the science of mind. There is nothing we know more intimately than conscious experience, but there is nothing that is harder to explain." What is the connection between mind and body? How does the brain create consciousness?

Chalmers calls this the "hard problem" of consciousness, distinguishing it from the "easy problems." He says that the easy problems include, among others, explaining the way we acquire information and focus our attention, and the difference between wakefulness and sleep. These are problems that can be solved through scientific inquiry. The problem of mind and brain, however, remains a mystery.

The question of the relationship between mind and body is an old one. For many centuries, philosophers have argued for different possible relationships between the two, and modern discussions of the topic are often derived from their age-old conclusions. Though the trend has shifted, the traditionally held view on the old mind-body problem has been the concept of *dualism*, the idea that there are two worlds: the physical and the mental.

During the seventeenth century, the mathematician and philosopher René Descartes decided that since he could imagine himself without a body, his mind must be nonphysical. The mind does not occupy space as physical things do, nor does it follow any of the laws that govern physical things. We cannot access the workings of someone else's mind; we can only make educated guesses about what the other person is thinking based on behavior. Consciousness, Descartes decided, must be somehow separate from the physical world. He believed that a person lives two simultaneous lives, one consisting of what happens to his body, the other of what happens in his mind. The first takes place in the physical world and the second in the mental world. In the mental world, I can imagine myself playing basketball in any way I choose. The mental basketball scenes affect my actions because I choose not to join the team. This shows that the mind can affect the physical world through its control of the body. These are the basic components of dualism.

Dualism is often illustrated by the analogy of the mind as a theater. The way a person typically perceives his own mind is that it is somewhere in his head looking out through his eyes, hearing through his ears, and so on. This is the basis for the theater view

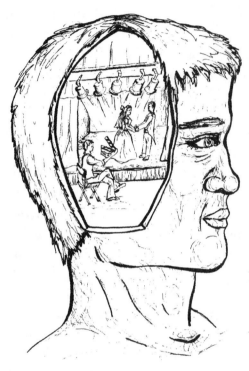

The theater of mind.
This dualist concept is commonly used to describe how consciousness seems to work.

of mind: a classic, centuries-old dualist concept. The analogy works as follows. Imagine that inside your head there is a stage. All the images you see, the sounds you hear, and all your other sensations are on the stage. The theater's spotlight represents where your attention is focused; anything inside the illuminated area is what you are most aware of. You are still somewhat aware of what is outside the spotlight, but you do not focus on it as actively. For example, suppose that you are looking at your friend's face from several feet away. Her entire body may be in your field of vision, but since the region around her cheeks, lips, and eyes are in the spotlight, this is the area in which your awareness focuses. You see that her eyes are green and that she is wearing makeup. You can detect that she is moving her eyes just slightly, and you are able to interpret a wince, a smirk, or a trembling of her lips, but her legs are outside the spotlight. You of course know that she is having a conversation with you while standing on her feet and wearing shoes, and may even remember those shiny black stilettos. To confirm these details, you would have to refocus the spotlight (your attention) on your friend's feet.

Backstage are the props and remnants of past productions

The homunculus. He or she is a tiny person in your head who represents your consciousness.

that you can look back upon to make decisions; they are the memories, experience, and knowledge you have acquired throughout your lifetime. And what about you? You are the director. You watch what is happening on the stage, using the spotlight to pick out areas of interest. You make judgments based on past productions. The director represents your *self*, your consciousness.

Philosophers often refer to the director as a *homunculus*, which literally means "tiny person." The homunculus (*homunculi* is the plural form) is the little man or woman inside your head who represents your consciousness. In illustrations, the homunculus is often shown inside the head, using buttons and levers to control the body. The homunculus, like the director, represents the self. Some prefer to use the words *soul* or *spirit*, but the implication is the same. All these terms really refer to one thing: your consciousness.

This dualist way of representing the mind has been maintained for so long that it has become incorporated into our language. For example, what does the pronoun "I" refer to? It refers to the conscious self. In the movie *Bicentennial Man*, Robin Williams plays a robot that wishes to be conscious. Since it is not conscious, it is programmed to refer to itself as "one" instead of "I." After receiving thanks for serving the family that owns it, the robot says things like "One is glad to be of service." Though the robot has a humanoid body, it lacks the human consciousness that "I" refers to.

The idea that a person is defined by his consciousness as opposed to his body is expressed in many other aspects of language. An example is our idea of death. When someone dies, he or she is said to "pass away" or "pass on." Since the person's body is still visible, these phrases must be referring to the person's conscious self, the soul.

In addition to the theater, there are several other metaphors used to represent the dualist mind-body relationship. One is that the body is a vehicle with each of us in the driver's seat. A similar analogy sees the mind as a pilot of a ship. Like the director, the driver and pilot are homunculi. However it is represented, the point is the same: there is consciousness within each of us that controls the body.

But how do we know this? How does someone know he is conscious? Descartes's famous answer is often quoted: "I think, therefore I am." Descartes was convinced that the fact that he thought he was conscious was evidence of his being conscious. After all, he couldn't have thought he was conscious without *thinking*. He had thoughts; therefore he must have been conscious.

This type of argument may appear strange at first. Since typical scientific methods cannot be used to come to conclusions about the mind, we must use a new kind of proof: a mental proof. In one dualist view, there is no way to study the mind scientifically. Consciousness is fundamentally a separate realm that we can learn about only through self-reflection, literature, poetry, and philosophy. This is the very reason why the humanities must remain separate from the sciences. But, other dualists hope that there can be a new kind of science for studying the mind, a science that accounts for the nonphysical nature of consciousness. Modern biology deals only with the interaction of physical things, like tissues, cells, proteins, and chemical compounds. Perhaps there will be a new science that will allow us to study the structure of thought or the rules of consciousness. However, toward that end, little progress has been made.

Some take the division between mind and body further to what might be called *hyperdualism*. Imagine that around the same time that the physical universe was formed, a second universe was created: a mental universe. The mental universe contained no matter, only a mass of floating consciousness. It was not organized into specific minds; it was a diffuse sea of consciousness. As countless years passed by, the physical and mental universes existed separately, one never causing disturbance in the other. Even as prim-

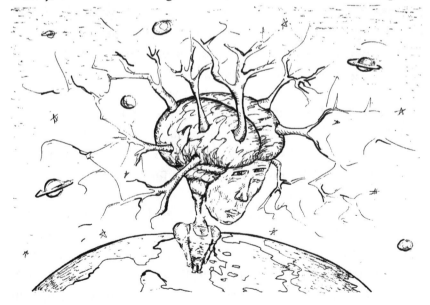

Hyperdualism. In this extreme view, the brain connects the universe of consciousness to the universe of matter.

itive life began to appear on Earth, the separation remained. The evolution of the brain changed all this. When the brain evolved, a link was formed between the physical and mental universes, thus creating conscious beings. Even today these two universes exist, providing us with both a bodily and a conscious existence.

This hyperdualist view is considered extreme by most modern philosophers. The customary view of dualism is that consciousness is nonphysical but exists here on Earth. Though separate from the physical world, it can interact with it through its control of the body.

If you accept a dualist view, then the answer posed by the title of this book must be no. We cannot simply be machines if we possess a nonphysical self that controls the body. My body can play basketball, but it requires my mind to control it. The *body* is a machine, but the driver is not.

Not everyone accepts dualism. Today, more and more people are rejecting it in favor of other theories. Philosopher Gilbert Ryle asserts that this belief in the "ghost in the machine" is mistaken. The ghost in the machine is yet another analogy similar to the driver in the vehicle, the pilot in the ship, and the director in the theater. It is absurd, he says, to think that there is some being inside of each person that controls his or her actions. It is equally absurd to say that someone has two separate lives. Ryle asserts that mind and body are parts of a whole; saying that there is a major division between them is a mistake.

Imagine that a foreigner is watching a baseball game for the first time and learns about each of the players. He is taught the functions of the batters, outfielders, pitchers, catchers, and umpire. He then says: "I understand what all these players do, but which one is responsible for team spirit?" What he does not understand is that team spirit is not a specific operation. It is an *idea* associated with the effective running of a baseball team. Similarly, consciousness is not a separate human operation. It is an idea associated with the proper functioning of a human being.

Here is another example. After the baseball game, the foreigner asks to see a university because he has never seen one before. His guide gives him a tour of the library, takes him through the individual college buildings, points out the dormitories and administration buildings, takes him to the student commons, and shows him the gym and sports fields. The foreigner is confused. He says: "I have seen the library and the classrooms and the gym and all the rest of it, but which building is the university?" The foreigner is making the same mistake as before. He thinks that the university is a separate unit from the buildings, when actually all the buildings combined *make up* the university. The university is not a building in itself; it is an idea.

In both examples, the foreigner confuses the meaning of the words he uses, *team spirit* and *university*. This is the same problem with the dualist view of consciousness, Ryle says. Dualists make the mistake of thinking that consciousness is a power and that it is separate from the body, when it is actually just an idea used to describe the mechanics of the brain. *Consciousness* is a word used to describe the way the brain works. It does not refer to some nonphysical force. To Ryle and others, dualism is absurd.

Some critics accuse dualism of being unrealistic. Most dualists believe that since the mind and body are separate, each one can exist on its own. When someone dies—when the ghost leaves the machine—his or her consciousness continues to exist. This implies that there are disembodied minds or spirits floating around. According to critics, dualists believe in ghosts.

Some believe that dualist views of the mind are just pointless. In the theater view of mind, the director perceives the world as if on a stage and makes decisions that control the body. Again, this analogy serves the same purpose as that of the driver to his vehicle and the pilot to his ship: a homunculus that represents consciousness interprets the senses and makes decisions. However, this scenario leaves out one element: how is the *homunculus* conscious? Is there another homunculus inside his head and yet another homunculus in that homunculus's head?

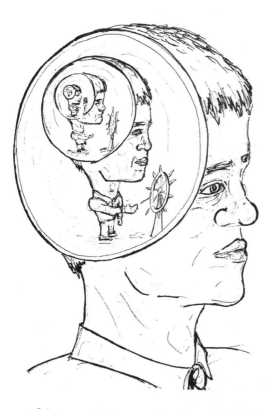

The homunculus problem. Critics say that believing you are conscious because you feel conscious doesn't solve anything.

This argument can also be used against Descartes's belief that thought is evidence enough for demonstrating his consciousness. "I think, therefore I am" is like saying "I am conscious because I feel the consciousness inside me." But how is *that* conscious? Critics say that Descartes really doesn't solve anything; he just avoids the explanation. Instead of explaining why he is conscious, he says that he feels the consciousness within him. This assertion results in the problem of an infinite series of homunculi, not a very good answer. The mystery remains.

Some philosophers maintain that the mystery will never be solved. The mind has limitations in understanding its own mechanisms. Maybe we are not yet advanced enough in our evolution to solve a problem like this. It is likely that animals have limits as to what they can understand. You don't see too many dogs around that could understand Einstein's relativity theory. Animal minds are *closed* to such advanced concepts. Our minds surely have boundaries; the understanding of consciousness may simply be beyond them.

Others advocate different mind-body relationships. *Idealism,* for example, is the idea that everything in the world is somehow a product of consciousness. According to idealism, everything we see, hear, and feel is a creation of the mental world and there may or may not be events in the physical world to which they correspond. This belief is held by few philosophers and will not be discussed in this book.

The relationship that poses the most serious challenge to dualism is *materialism.* Ryle's argument against dualism comes from a materialist standpoint. Materialists hold that consciousness can be entirely explained as a physical phenomenon; it is a matter of brain mechanics. The fact that it has remained a mystery for so long is irrelevant. Years ago, many people thought that the way genetic characteristics are passed on from generation to generation would never be understood; then DNA was discovered and the mystery was solved. A new discovery will solve the consciousness mystery as well.

It is important to note that dualism is still a strongly supported opinion. The arguments against it have not eliminated it as an approach to the mind-body problem. Though materialism strongly challenges dualism, it has by no means overturned it.

Still, because of innovations in science and technology, many people today accept a materialist, rather than the classic dualist, view of the mind. With the help of new discoveries, why shouldn't we be able to find a physical explanation for consciousness? According to materialists, the fact that someone can imagine a basketball game does not mean that consciousness is separate from the body. The brain somehow creates consciousness and, as science progresses, we will understand how—just as we understand other bodily processes.

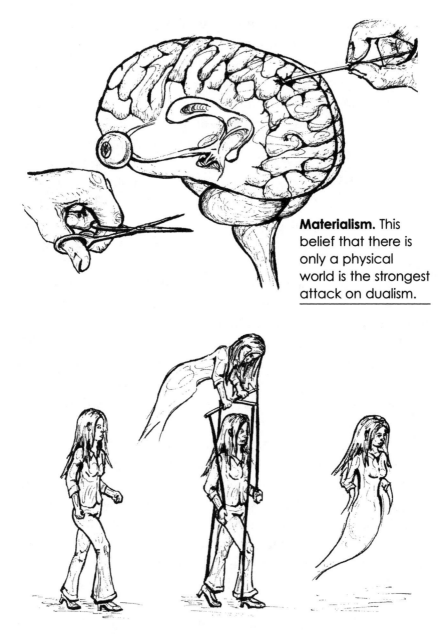

Materialism. This belief that there is only a physical world is the strongest attack on dualism.

The spectrum of opinion. These are the main categories of philosophical views on the mind-body problem: materialism, dualism, and idealism.

FURTHER READINGS

The comparison between the "hard" and "easy" problems of consciousness comes from *The Conscious Mind* by David Chalmers. His quote in the beginning of the chapter is from the introduction to his essay "Facing Up to the Problem of Consciousness" published in the *Journal of Consciousness Studies*, pp. 200–19. To learn more about the theater of the mind, look for *In the Theater of Consciousness* (1997) by Bernard Baars. References to the homunculus can be found in many books, including *In Defense of Human Consciousness* (1997) by Joseph Rychlak and *The Rediscovery of the Mind* by John Searle. The concept of hyperdualism is described in *The Mysterious Flame* by Colin McGinn. McGinn is a philosopher who doubts that we will ever understand consciousness because of our limited cognitive abilities. Gilbert Ryle's attack on dualism as well as a good summary of dualism as it is attributed to Descartes can be found in his *The Concept of Mind* (1949). The examples of the foreigner viewing baseball and the university were taken from here as well.

4
THE MECHANICS OF MIND

My younger brother Benny and I went to a Chinese restaurant with strict instructions that he would have ten dollars to spend on his meal and would have to eat at least four pieces of broccoli. Benny sat down and opened the menu. Should he order egg rolls or egg drop soup with his noodles? What about some Kung Pao chicken? Dessert? After twenty minutes or so, the order arrived—with a side of fresh broccoli. Benny cringed. He never liked broccoli. I assured him that the broccoli would taste good and that it would be good for him. Benny sighed and put the green vegetable in his mouth. He wrinkled his nose, groaned, and spat out the broccoli, but finally promised to eat it later. We dug into our noodles, Benny fumbling with the chopsticks. After the fortune cookies, I noticed with pleasant surprise that the broccoli was missing from Benny's plate—but I was suspicious. Soon enough, the evidence appeared: concealed within a crumpled napkin were four pieces of broccoli.

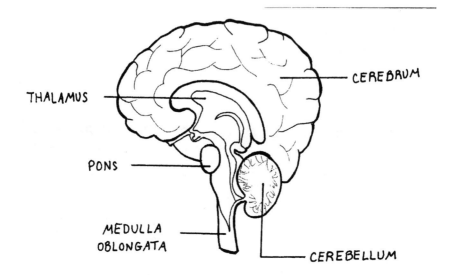

Parts of the brain. The areas of the brain have different general functions, but they constantly communicate with each other and function as a unit.

The above scenario demonstrates a number of the powers of consciousness, all of which, according to some scientists, can be attributed to the mechanical workings of the brain. These scientists assert that the brain is a machine—a biological machine—but a machine nonetheless, just as the heart and stomach are machines. Every conscious power that my brother demonstrates in the restaurant has a mechanical explanation, whether or not we understand it—and we do understand a lot.

When my brother sees broccoli being offered to him, a chain of chemical reactions occurs in his brain. Billions of electrical signals are zipping throughout his nervous system. They start in the sensory organs that register information from the environment and stream along nerve fibers to the brain, which interprets them and sends out instructions that control Benny's response. These signals are conducted by a network of messenger cells called *neurons*.

Neurons (bundles of which are called "nerves") have three

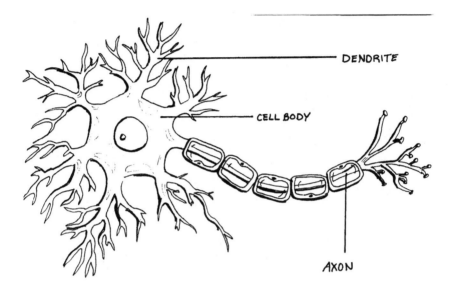

The neuron. This is the basic messenger unit responsible for sending and receiving electrical signals.

major parts: a cell body, dendrites, and an axon. The cell body nourishes the neuron and allows it to grow. The dendrites are thin, branching fibers that receive incoming signals from other neurons. The axon is a single fiber that sends out signals for other dendrites to receive.

Neurons are grouped into *circuits* that allow the electrical signals to flow from one neuron to many, from many neurons to one, or in a circular pattern. When a quick response is needed, a single large neuron can cause thousands of muscle fibers to contract. When the challenge is more complex and more consideration is needed, the network allows for signals from many processing centers to converge, generating a single response. Scientists assert that this network of tiny interactions forms the basis for consciousness.

Though this network is spread over the entire organization of the brain, the specific human ability for higher-level thought is centered in the *cerebrum*. Making up three-fourths of the entire

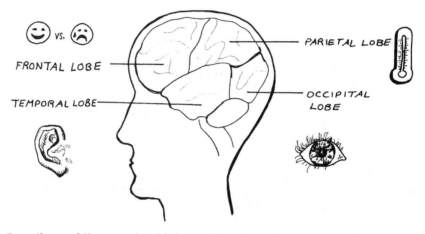

Functions of the cerebral lobes. Using imaging technology, scientists have been able to map the functions of the cerebral regions.

brain, the cerebrum manages the senses, voluntary movement, emotions, language, and reasoning. Scientists have discovered the different regions, or *lobes*, of the cerebrum in which the various powers occur. They have accomplished this by connecting electrodes from various points on a subject's cerebrum to a computer that maps out the areas that are activated. For example, when the subject walks around, the top of the brain is highlighted on the computer screen. When the subject listens to music, the sides of the brain are highlighted on the screen. So, scientists not only know how neurons interact in the brain, they also know *where* they interact—they can trace senses, emotion, behavior, and speech to their cerebral sources.

Let's now return to Benny's meal. When my brother sees the broccoli offered to him, what happens? First, his eyes, like a camera, focus on the vegetable. Next, his retinas convert the visual image into a pattern of electrical signals that travels along a bundle of neurons, called the "optic nerve," to his *thalamus*. The thalamus acts as the body's main sensory switchboard—it receives signals from our many sensory receptors and directs them toward

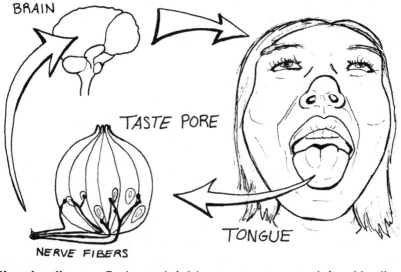

BRAIN

TASTE PORE

TONGUE

NERVE FIBERS

Signal pathways. Taste and sight are processes explained by the signaling network of the nervous system.

the appropriate parts of the brain. In the case of sight, when the signals from Benny's eyes reach the thalamus, they are redirected to the corresponding visual area of his cerebrum (the visual cortex). There, groups of neurons interpret the pattern of signals in order to create a meaningful picture of the world.

To recognize the object in that picture as broccoli, Benny needs two other mental components. The first is memory and the second is the ability to make comparisons. The first time Benny saw broccoli, the pattern of signals from his eyes was stored. Now when he looks at broccoli, he can recognize it. This is because the signals that reach the visual cortex of the cerebrum also pass through nearby "association areas" that compare the signal pattern just received with patterns stored in memory.

The taste of the broccoli is processed in a similar way. The taste buds on the tongue send signals to the thalamus, which redirects them to the primary taste cortex of the cerebrum and its surrounding taste association areas. The disgust response arises from the temporal lobe of the cerebrum. The signal patterns are then processed in the frontal lobe, which determines the spit response. The processes of sight and taste have mechanical explanations, as do the processes of comparison and recognition.

The extensive neurological research that has helped us to understand whatever we do know about the brain (and there is far more research going on than I can hope to summarize) seems to show that Benny is controlled by biological mechanisms. The very fact that Benny is hungry in the first place, and later full, is controlled by the *hypothalamus,* a small part of the brain that manages the body's internal environment by controlling blood pressure, body temperature, and the chemicals released in the body. When someone is in need of nutrients, the hypothalamus triggers the release of chemicals in the body that make the person feel hungry. When enough food has been taken in, other chemicals are released that cause the person to feel full. The nutrients from the food we eat are then distributed among body cells by mechanical processes. All of this is controlled by our biological makeup. How far does this mechanical control go?

Our ability to have emotions is something that people are tempted to set aside from other aspects of consciousness because it seems to be so much a part of our identity. At first glance, emotions do not seem to be in the same category of mechanical processes like tasting or seeing or digestion. However, scientific research shows that emotions also depend on the mechanics of the nervous system. The main evidence for this comes from experiments in which various parts of the brain are stimulated with electricity. Exciting one part of the temporal lobe causes the patient to experience intense fear; doing so to other parts causes feelings of loneliness, sorrow, and, yes, even disgust.

Even if a patient has countless problems, a joyous feeling that everything will be fine can be brought about by appropriate stimulation of the brain. In one experiment, the patient was a severely depressed man. He could barely hold back tears as he described how he felt responsible for his father's illness. When the proper part of his brain was stimulated, however, he suddenly ended the emotional speech. Within a few seconds, he grinned widely and began talking about an upcoming date with his girlfriend.

The cerebrum receives information from all over the body, processes the information to form a meaningful picture of the world, and organizes the behavior by which to respond. Patterns of response, such as running from danger, are stored in organizations of neurons and can be changed in the cerebrum. An aggressive response, for example, can be stored in a different group of neurons. In experiments, scientists can cause animals to behave in a hostile way by exciting these neuronal groups. When a pigeon alone in a cage is stimulated in this way it behaves angrily, as if there were another bird in its territory. The pigeon walks threateningly around the invisible bird, preparing to attack it. Biological mechanisms seem to control its behavior.

Maybe Benny is the same way. We have seen how scientific findings explain how Benny learns and remembers, how he recognizes the appearance and taste of broccoli, and how he responds with disgust. His emotions and behaviors have mechanical explanations. Does this mean that his consciousness is mechanical? If it is, then Benny is a machine—a biological machine. However, we can't be sure, because there are some events that happened in the restaurant that biology cannot explain. For example, we still don't know how the conscious experience of tasting broccoli occurs. How do electrical signals cause us to taste broccoli as we do? How does Benny *experience* the taste of the vegetable? Why do these signal pathways cause the broccoli to taste bad to my brother and not to me? Scientists cannot say.

Benny's biological makeup cannot explain his strategic approach to avoid eating the broccoli. He could have argued about his having to eat four pieces of broccoli, but after thinking it through, he decided that it would do no good. He concluded that it would be easier to try to fool me by pretending to eat the broccoli while actually hiding it in a napkin. How did he know how to hide it effectively?

Similarly, the way Benny orders his meals is beyond any known biological mechanisms. Before he orders his food, he decides what kinds of food he was craving and how much he was willing to sacrifice quality for quantity. He likes egg rolls more than egg drop soup, but egg rolls are more expensive. Is he in the mood for Kung Pao chicken? Is it more filling than egg rolls? Does he want dessert today? He has never had many of the foods on the menu, but he can easily imagine what they will taste like. His planning to create his ideal meal demonstrates higher-level thought that is beyond known brain mechanics.

I also demonstrate this advanced thought at the table when I search for broccoli that Benny may have hidden. Suspecting that he may have tried to cheat me, I investigate by searching through the things on Benny's side of the table. I find the broccoli and come to the conclusion that Benny hid it in order to avoid eating it. Here I demonstrate mental abilities that, once again, brain researchers so far are unable to explain.

Though neuroscientists have a lot to say about how the brain works (I have provided an extremely simplified account), they will tell you that they actually know very little. The knowledge they do have about the brain deals mostly with what David Chalmers would call the "easy problems" of consciousness. We are far from knowing how the brain creates consciousness. From the biological standpoint, we do not have enough evidence to prove that Benny is a machine; we have only theories and possibilities.

FURTHER READINGS

Much of the technical information in this chapter can be found in *Anatomy and Physiology* by Arthur Guyton and in *The Oxford Guide to the Mind*, a collection of articles compiled by Geoffrey Underwood. The experiments I discuss in this chapter are found there as well. Another book to check out is *Consciousness* by J. Allan Hobson. It is beautifully illustrated with descriptions of the biology of sleep, wakefulness, and other brain states.

5

CONSCIOUSNESS EMERGES

How is it possible for nonconscious material in the brain to generate consciousness? How can biological mechanisms create a homunculus or self? How can they create free will? This problem has troubled philosophers as well as neuroscientists for some time. To develop a biological theory of consciousness, these questions must be answered.

Scientist Francis Crick writes that it is possible for consciousness to emerge from physical structures without resorting to dualism, the idea that there is a mental world separate from the physical world. No stranger to scientific mysteries, Crick was awarded the 1962 Nobel Prize for Physiology or Medicine, along with James Watson, for discovering the structure of DNA. The soul, he says, is a myth that has existed for so long only because science has not been ready to attack the problem of consciousness until now. Years ago, nearly everyone believed the myth that the world was flat. The fact that the majority of people believe something doesn't make it true.

Before Galileo and Newton, it was thought reasonable that the movement of the planets in the solar system required the guidance of angels. Before Charles Darwin's theory of evolution, there was no way to explain how complex living beings could have come about without the intervention of an omnipotent Creator. Crick says that the myth of a nonphysical soul that survives the body after death is a similar symptom of insufficient knowledge of science. He asserts that the emergence of consciousness from purely physical structures is possible, even if the mechanism is not yet understood.

The reason one might doubt this possibility, Crick says, is that the individual structures in the brain—such as neurons—have none of the properties of consciousness. How, then, could many of these structures working together generate it? Crick notes that there are many phenomena in nature that are more than the sum of their parts. For example, compounds in chemistry, such as salt, have completely different properties than the elements they are made of. Typical table salt (sodium chloride) is made of the elements chlorine and sodium. Chlorine gas is a deadly poison, but sodium chloride is an essential nutrient. The compound containing sodium and chlorine has completely different properties from sodium and chlorine, but that doesn't mean that the compound has a mystical, salty spirit.

Life itself is another example. All of us are, at the most minute level, made out of nonliving, subatomic particles. The living cells in our bodies are composed of nonliving atoms, which are composed of nonliving protons, neutrons, and electrons, all of which are composed of nonliving quarks. If nonliving, almost invisible particles can be the building blocks of complex, living beings, it follows that nonconscious structures like neurons can be the building blocks of consciousness.

In the previous chapter we looked at the basic signaling mechanisms of the brain. We saw how neurons throughout the body can receive sensory information from the surrounding world and transmit it to the brain for processing. We also saw that specific brain areas seem to be the source of emotions as well as simple behaviors. How might all this information be developed into a biological theory of consciousness?

Scientist Gerald Edelman puts forth one possibility. Edelman suggests that we consider neurons in groups that work together. He asks us to imagine sheets of neurons that correspond to sheets of receptor cells, such as an area of skin or the retina of the eye. These sheets are in constant communication with each other. For example, the visual cortex of the brain might have thirty or so of these neuronal groups constantly interacting.

Edelman hypothesizes that the relationship between these neuronal groups provides a system for dividing things in the world into categories and representing them in a unified way.

The mountain of memory. Edelman uses this metaphor to describe the evolving nature of memory.

Suppose, for example, that you are looking at a person's face. Each group of neurons is responsible for processing a certain aspect of what you see. Some will process the outlines of facial features. Others will process colors, and still others will process movements. Rapid interactions between the groups put everything together and create a unified representation of the world. They enable us to create categories of objects with similar properties such as cars, trees, or buildings. Whenever we come across objects, the categories we have created are modified, reinforced, and expanded. This perceptual categorization is a primitive cognitive ability from which, with the help of other simple abilities, consciousness might arise. Edelman insists that this ability to categorize is necessary but not sufficient for consciousness. In fact, a computer simulation of a robot has been able to perform the task. This ability is, however, a requirement for what Edelman calls "primary consciousness."

Primary consciousness involves the ability to have simple sensations and sensory experiences. It is a basic form of consciousness that is seen in animals that have brain structures similar to ours. Edelman distinguishes this basic consciousness from the "higher-order consciousness" that human beings also possess. Primary consciousness is the more basic, so we will address it first.

The capacity to separate things in the world into categories is only one aspect of primary consciousness. Memory is another. However, by memory, Edelman does not mean the simple storage of information, but an active process of modifying and reinventing existing categories. Imagine, for example, that you see a strange-looking car for the first time. It has spikes jutting out from the roof and razor-sharp teeth in front (perhaps it belongs to Godzilla). Although it is unlike any car you have ever seen before, you still recognize it as a car. The category of cars is recalled from your brain, compared to this car, and then expanded to include this new type of vehicle.

Edelman provides the following metaphor for how this process might work. Picture a mountain with a glacier near its peak. When the weather warms, streams flow into ponds and puddles along the side of the mountain. They represent stored memory—categories of stored information. The collections of water, like memory, are constantly changing shape as new streams (which represent information) begin to form because of changing conditions. The streams twist and turn and merge with other streams. The ponds become linked to one another as the stream meanders over the surface of the mountain.

Memory is fluid, according to Edelman, not a simple historical record like symbols scratched in rock. Our recollection of events can fade and intensify. Streams of memories run throughout the mind, sometimes fading, sometimes surging, sometimes running into one another. People, places, events, and ideas can change, separate, and converge as we go through life recalling them as memories.

Our memories are shaped by our perceptions, which in turn are shaped by our memories. For example, suppose three people standing together witness a robbery being committed. Ten years later they are interviewed separately. Each of them is likely to tell a slightly different story. One of the three might recall having heard shots fired in the bank, while the others have no such recollection. One of them might remember the gruffness of the ringleader's voice. Another might be able to describe the kind of clothes the robbers were wearing. Each witness will remember certain details clearly, some not as well, and still others not at all. The same event will be recalled differently by each one. Whether the witnesses realize it or not, their memories undergo revision and embellishment over time. Edelman stresses that this evolving nature of memory is crucial for consciousness to arise.

In addition to memory, Edelman says that a system of learning is necessary for primary consciousness. Learning adds an important aspect of consciousness: the preference of some experiences over others. Consider a person strapped to a chair. In front of the chair is a console with two buttons, one red and the other yellow. When the red button is pressed, a massaging feature is activated in the chair. When the yellow button is pressed, the chair delivers a painful shock. Once the effects of pressing each button are stored in the person's memory, the person seated in the chair will *learn* to value one over the other (I'm assuming the massage, but you never know). In the future, unless you try to be funny by switching the buttons around, the person will press only the button that he or she prefers. This simple kind of learning is essential for primary consciousness.

Edelman goes on to say that there are two other requirements for primary consciousness. First, one must be able to distinguish the self from the nonself. Edelman is not yet referring to the human sense of self (the homunculus). That is part of higher-order consciousness. Here, he simply means that an organism must realize that it is an entity separate from the outside world. This might involve, for example, having some basic urges such as hunger. When an organism feels hungry, it will find food and consume it. This gives the organism a basic sense of itself (it *feels* hungry) and creates a separation between the organism and other objects, such as the food.

Edelman believes that the final requirement for primary consciousness is having a sense of time. Being able to make sense of sequences of events (A occurring before B; C taking place before D; etc.) is critical for conscious activity. Edelman guesses that the brain processes that make this possible are similar to and connected with the processes for forming categories and creating concepts.

So, to recap, Edelman believes that to have consciousness, an organism must have a number of abilities. It must be able to categorize objects in the world in order to form concepts of what various things are. For example, it should be able to place sheep in a different category from porcupines. It must also be able to store these categories in memory, which is an active process of creation rather than a simple storage mechanism. So, if the organism sees a porcupine that is unusually obese, it has to expand its category of porcupines to include the unfamiliar form. The organism also needs a system of learning that allows it not only to acquire knowledge, but also to distinguish what it likes from what it does not like. For instance, the organism may like to pet the sheep but not the overweight porcupine. Finally, the organism must be able to make a distinction between itself and the outside world and have a sense of time. Therefore, when the ridiculously fat porcupine walks by from left to right, the organism should realize that the porcupine is a separate entity and that the porcupine was on the left first and on the right later. According to Edelman, the interaction between all of these systems is primary consciousness.

The communication between groups of neurons and their related cells and chemicals makes primary consciousness possible. Primary consciousness, in turn, makes the more advanced "higher-order consciousness" possible. According to Edelman, this sophisticated level of consciousness arises only when human beings experience social interaction. This kind of interaction, which requires language, contributes to the development of human thought and decision making. Through our use of language and social contact with one another we develop our own human identities. In Edelman's theory, this ability, along with all the aspects of primary consciousness, constitutes the human mind.

Now that we have outlined a theory, let's see if we can explain the two aspects of consciousness cited at the beginning of the chapter: the self and free will. We will start with the self.

We know that the brain has many regions that serve distinct purposes. According to Edelman, neurons work in separate groups that focus on certain tasks. Despite all the billions of interactions going on separately in the brain, it seems to us to be a central controlling power. Our experiences have a regular perspective. It is as if a homunculus exists in each of our heads, monitoring what goes on. How does this controlling self arise from so many separate biological activities?

Neuroscientist Antonio Damasio attempts an answer. He says that there are three aspects of the self. The first component is history. Key events in our lives are stored in memory and can be recalled. These experiences help form our identities as well as the perspectives from which we view the world. Since, as Edelman says, memory is not a static process, everything we experience expands upon and reinvents what we already know. Your point of view is constantly being constructed.

So, your identity is formed by the facts about yourself that are stored in memory. These can include your name, where you live, who your parents and siblings are, your goals, your fears, and anything else that contributes to your perspective. You perceive the world through the lens of your life experience and knowledge.

Another of Damasio's components of the self is having knowledge of one's own body. This idea is similar to one of Edelman's requirements for primary consciousness: the ability to distinguish self from nonself. Even if you were born in a place without mirrors, you would still sense that you have a body. You would know how to use your arms and legs and you would figure out how changes in your body affect your mind. The awareness you have of your body also contributes to your identity because it allows you to feel that you are an individual. You would recognize that others in your world are not you. They exist and move independently of you.

A third component of the self, according to Damasio, is the capacity for language. This is not to say that you must be able to speak a nationally recognized language such as English or Portuguese to have a self. You do, however, need to develop some form of symbolic representation. In order to have a conscious self, you must be able to articulate the world around you, your experiences, and aspects of your body in a meaningful way. Otherwise you aren't *thinking*. As I write this, I am thinking in English. You might also be thinking in English as you read. However, you could also be thinking in French or Japanese, in images or in music or in symbols. Any system of expression will do. Many scientists and philosophers agree on the point that we would not be able to think without language. Just having a body and a history does not mean that you can attach meaning to the world. You need some form of language to do so. The capacity for language allows for the realization and expression of one's consciousness, the self.

Our ability to comprehend language is located in the left hemisphere of the brain. Though I do not specifically mention them, there are biological mechanisms responsible for all the other components of the self. Interestingly enough, not all of the brain is necessary for one to have a self. There is an operation called "split-brain surgery" during which the nerve fibers between the two hemispheres of the brain are cut. In some cases, the patient is left with two minds, *two selves*. The patient's left and

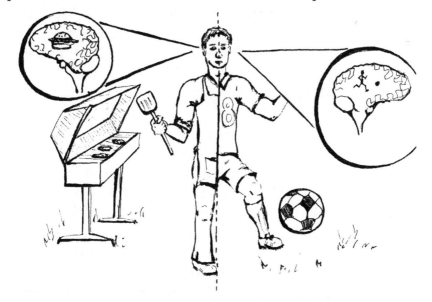

Split-brain surgery. Patients that undergo this procedure are sometimes left with two minds that control opposite sides of the body.

right brain hemispheres will operate independently rather than one hemisphere being subordinate to the other, as is the case when the hemispheres are connected. In addition, since each eye is controlled by a separate hemisphere, the patient's right side will be unaware of what the left side sees and vice versa. The left and right selves might independently initiate activities that conflict. For example, the patient might begin to button his shirt with one hand while trying to unbutton it with the other hand. From this and other evidence, many scientists agree that interactions between brain structures generate the self.

We see now how science might explain the biological origin of the self. However, the question remains: how does science explain the control that the self has over the body? In other words, what is the biological system that drives free will? For one possible answer, we return to Francis Crick.

There are parts of one's brain responsible for making plans and formulating decisions. These plans and decisions are the results of countless calculations performed by the mechanisms in his brain. The person, however, is unaware of these calculations. It follows that he is aware of the results of his brain's unconscious calculations but ignorant of the calculations themselves. The person knows the decision he makes but not what his brain does to come to that decision.

So, in Crick's view, it appears to this person that he has free will, but in reality he does not. The plans that he comes up with are the result of unconscious brain calculations; the decisions that he makes are generated in the same way. The actual causes of the decisions may be incredibly difficult to identify because a small event may cause a significant change in the final stages of the brain's calculations. Since there are an enormous number of factors that can affect these unconscious calculations, and since each of the factors has a result that is stored and then entered into other calculations, the final result is practically unpredictable to the ordinary observer. This might give someone—keeping in mind that we are all only ordinary observers of the workings of our own brains—the wrong impression that his will is actually free because the final result of these calculations *appears* unpredictable. However, Crick says, it is actually those unconscious calculations that caused the action. What appears to be freedom is the unconscious mechanical outcome of all these prior calculations.

Crick claims that scientists have discovered the source of what seems to be free will by studying patients with certain types of brain damage. He cites a description given by Antonio Damasio (the scientist discussed earlier) of a patient with an unusual problem. After suffering the brain damage, this woman appeared alert but did not respond to anyone around her. When doctors asked her questions, she would follow people with her eyes and act as if she understood the questions posed to her, but she would not speak to anyone.

A month later she was mostly recovered. When asked why she would not respond to anyone before, she said that she understood the conversations going on around her but had "nothing to say." She felt as if her mind were "empty." Crick interpreted this to mean that she had lost her will. The question then brought to mind was: where in her brain did the damage occur? The damage was found to be in an area called the "anterior cingulate sulcus" (sure, I guessed that too), an area that receives and processes signals from many sensory areas of the body. Crick suspects that this area is the seat of the will and that the above condition is evidence.

Crick considers the following disorder to be further evidence. There is a condition known as "alien hand syndrome" where a patient's hands might act against the patient's will. The left hand, for example, might suddenly grip a nearby object without the person deciding to do so. Here, again, the will appears to be affected. Sometimes the patient is unable to cause the hand to release the object and may have to use the other hand to remove it. One patient found that he could make his "alien" hand let go only by screaming at it. The damage in this case was also in the

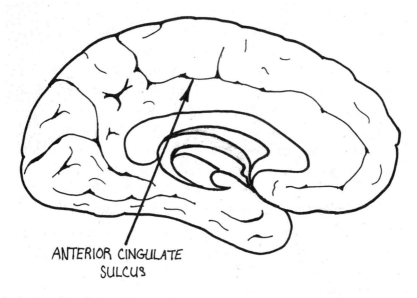

ANTERIOR CINGULATE
SULCUS

Anterior cingulate sulcus. Crick believes that human free will arises from the operation of this area of the brain.

anterior cingulate sulcus. Crick believes that this confirms that this area of the brain is the seat of the will.

Crick believes that this is the part of the brain that generates what we think of as the will and that the brain unconsciously calculates our decisions for us. His conclusion is that free will does not exist, that we don't have free control over our lives. The brain is programmed to make calculations for decisions. We think that those decisions were made by our free will, but they were actually made by the brain. According to Crick, our physical bodies are in control. There is no self or homunculus. There is only the brain.

If correct, Crick has shown that you are a machine. That you are made out of organic material instead of metal does not matter. What does matter is that your humanity is purely the result of interacting physical components. It would be consistent with Crick's ideas to say that if a mad scientist were to know everything there is to know about your brain and body and had sufficient technology, he *would* be able to predict what you will do and think. You are an organic machine.

Crick and Edelman believe that consciousness emerges when all the physical parts are organized properly. *There is no soul.* In their view, when someone's physical body dies, the mind dies with it. The brain alone generates the mind.

Crick and Edelman have presented speculations, not scientific findings. Science is not yet able to confirm their theories, and is thus unable to tell us with certainty whether or not we are organic machines. But, if we could start with physical materials and construct an entity with consciousness, *that* might prove that we are machines. Could we build a mind?

FURTHER READINGS

To learn more about Francis Crick's work on consciousness, look for his book *The Astonishing Hypothesis: The Scientific Search for the Soul.* There are also interesting points in a paper he wrote with Christof Koch called "Consciousness and Neuroscience." It was published in the journal *Cerebral Cortex.* Gerald Edelman has written a number of books on consciousness, including *The Remembered Present: A Biological Theory of Consciousness* and *Bright Air, Brilliant Fire: On the Matter of the Mind.* A more recent publication of his is *A Universe of Consciousness: How Matter Becomes Imagination.* The metaphor of memory as streams on a mountain was taken from that book. Some of the information on how the self might arise was taken from Antonio Damasio's book *Descartes' Error: Emotion, Reason and the Human Brain.* Another book to check out is *Consciousness: A User's Guide* by Adam Zeman. For a good discussion of some biological theories of consciousness, look for philosopher John Searle's book *The Mystery of Consciousness.*

6

HOW TO BUILD A MIND

Suppose that, while you are sleeping, a scientist makes a living copy of you. Using various materials, he copies every aspect of your body and brain in order to create a perfect, conscious duplicate. Let's call your copy Replica (we will assign to Replica the pronoun "it"). Replica has all of your memories and is therefore convinced that it is you. Your friends and relatives will be certain that Replica is you because Replica knows everything you do. When Replica is created, your consciousness does not move or change. In fact, since you have been sleeping, you have no idea that you were copied. When you are told, you would probably say that Replica is not you, but someone else. You would certainly object to the idea that, now that Replica has been created, you are no longer needed. To you, Replica would be a separate person.

Let's consider a second thought experiment, which is, as the term implies, an experiment we do in our heads. Suppose that the scientist replaces a small part of your brain with material that serves the same functions as the removed chunk. You don't feel any different; your consciousness is unharmed. There are people like this today who have implants that help with hearing problems and Parkinson's disease. Suppose now that the scientist replaces a little more of your brain, then a little more. Gradually, he replaces your entire body with other parts. At the end of the operation, you still feel the same. All your parts are working properly and it seems to you that your identity has been preserved. You feel that you are the same person now that you were before the procedure.

At the end of the operation in this thought experiment you have become the same as Replica from the first thought experi-

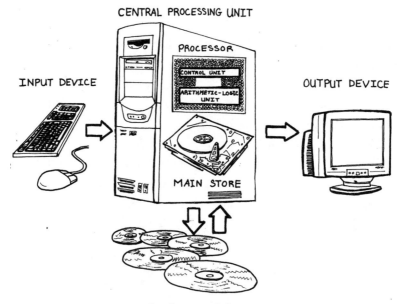

The Basic Movement of Information in a Computer. An input device, such as a keyboard, sends information to be processed. The results are displayed by an output device, such as a monitor. Information can be stored on the hard drive (main store) or on disks (backing store).

ment. We have, however, established that Replica is not you, but someone else. Therefore, in the second scenario, after all the particles in your body are replaced, you have become someone else. But isn't this what happens naturally? Cells in our bodies continually die and are replaced by others. Most particles in the body are replaced every few months. Are we therefore constantly being replaced by someone else?

To answer, we must first understand what we already are. Computer scientist and inventor Ray Kurzweil says that we can't just be a set of particles because, since those particles are constantly replaced, that would imply that we are being constantly replaced. He doesn't think that is what happens. Rather, we are an *organization* of particles—a pattern of materials and energy. Though the particles that make up our structure may change, our organization remains the same.

Since each of us is a pattern, Kurzweil says, that pattern can be scanned and duplicated. Each of us can be copied and nobody will be able to tell the difference between our copies and us. Kurzweil predicts that by the year 2030 we will be able to upload the mind into a computer and reconstruct its organization and operations. He says that this will lead to the building of computers with all the powers of the human brain—computers that are conscious.

If it is possible to create a machine with our conscious powers, it is likely that the technology used will be some sort of an expansion of what we already know. Today, a computer is capable of four things. First, it can accept instructions (input), such as the command to add the numbers two and three from an input device such as a keyboard. Second, the computer is able to process the instructions by adding the numbers. Third, it can return a result (output), the number five in this case, displaying it on the output device (such as a monitor). Finally, the computer can store results in a memory device such as a hard drive or a disk.

The set of instructions received by a computer is called the

```
Program: John Detector;

Begin

  Write ('What is your name?');
  Read (name);
  If name = 'John'
     Write ('Your name is John');
  Else
     Write ('Your name is not John');

End.
```

John Detector Program.
This simple program will prompt the user for his name, check to see if it is John and then state whether it is.

program. These instructions cause electricity to surge along a complex circuit pathway that, in turn, leads to a response of some sort. For example, given a user's name as input, a program could figure out whether the user's name is "John."

Computers translate the instructions in the program into *binary* code, which uses two symbols, usually represented by zeros and ones, to manage the opening and closing of circuits in the computer. For example, if part of the program translates into 100111, the computer will connect one circuit, allowing electricity to flow through that region, disconnect the next two circuits and connect three others. The opening and closing of computer circuits allows the electricity to flow down an enormous variety of pathways, creating a vast number of possible responses.

The human system of neuron signaling is often compared to this binary system of the computer. Neurons either signal or don't signal; circuits are either connected or disconnected. The similarities between the networks of neurons in the brain and circuits lead many to believe that the brain is a computer and the mind is a computer program. If this is true, it not only means that it will be possible to build conscious machines, but also that we are machines.

The system for detecting whether the user's name is "John" is an example of an ordinary program and doesn't seem at all as powerful as what the brain can do. Certain types of advanced programs are said to provide computers with "artificial intelligence" (AI): the ability to apply a set of rules, created by a programmer, to solve problems in a way that gives the machine an appearance of intelligence. These are the programs that, some say, rival the power of consciousness.

A well-known example of the power of AI is the chess-playing computer Deep Blue, created by IBM. The game of chess is vastly complex. There are an incredible number of possibilities to consider before making each move. A player has up to 218 moves at each turn. It is estimated that there are between 10^{43} and 10^{50} possible arrangements for pieces on a chessboard—quite a bit to think about. To beat world chess champion Gary Kasparov in

Deep Blue and Gary Kasparov. The Deep Blue computer processing system weighs 1.4 tons and is programmed to do nothing but play chess. It gained fame by defeating world champion Gary Kasparov (left) in 1997. (AP Photo/Adam Nadel)

1997, Deep Blue used exceptionally fast processors in order to test two hundred million positions per second while Kasparov could test about three. Deep Blue not only demonstrates the functional power of AI (in chess at least), but also that AI can be "smarter" than people are.

As powerful a machine as Deep Blue is, it is still referred to by some as having "Weak AI," the ability to solve limited, though complex, problems. Machines with Weak AI are not conscious. A far more ambitious concept is "Strong AI." Proponents of Strong AI claim that properly programmed computers can generate consciousness. Can we build on what computers can already do and build a machine that possesses Strong AI?

There are three general approaches that researchers might take to use computer power to build a mind. One way is to take on the incredibly tedious task of programming a machine with all the rules of various processes. The John Detector program has one rule: the user's name is John only if he enters the name "John" (what insight!). It's hard to imagine the immense number of rules it would take to explain the many facets of a typical conscious life. Nevertheless, there are those who are trying to accomplish this task.

Since 1984, programmers have been working on CYC (pronounced like the second syllable in "encyclopedia"), a program designed to apply a humanlike system of common sense to solve problems. The CYC program has over a million basic rules. Here are some examples quoted from the project's director:

- You have to be awake to eat.
- You can usually see people's noses, but not their hearts.
- You cannot remember events that have not happened yet.
- If you cut a lump of peanut butter in half, each half is also a lump of peanut butter; but if you cut a table in half, neither half is a table.

The leader of the project says that CYC has made impressive inferences using its million or so rules as a basis. In one test, CYC was given the following statement to analyze: "Huck thought it felt good to rest his bare toes on the warm rough wooden sock." Even though it is very likely that the words *toes* and *sock* would appear in the same sentence, CYC guessed that the user accidentally hit *s* instead of *d* on the keyboard and really meant to type *dock*. Those in charge of the CYC project believe the system will have many applications. Maybe consciousness will be one of them.

A second approach to machine consciousness is to duplicate the intricate neural structure of the brain. To accomplish this, scientists must first analyze how the neurons in the brain are connected. Ray Kurzweil says that this process has already begun. Several human brains have been thinly sliced (presumably after their owners were no longer around) and analyzed by researchers who then saved the structural information they discovered in a giant computer database. However, this information is nowhere near enough. To learn more, Kurzweil says, we will send micro-

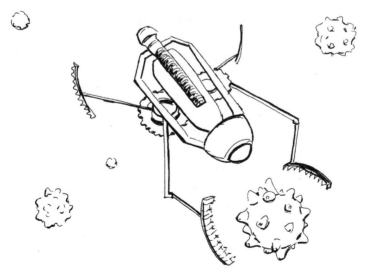

Nanobots. Kurzweil says that these microscopic robots will one day be used to construct a highly detailed map of the brain.

scopic robots called "nanobots" swarming throughout the body, programmed to record all the brain's connections and activities. The next step will be to create what are called neural networks, or "neural nets," mathematical models of neurons that will interact in a comparable way.

In discussing artificial intelligence, it is often asked whether a computer could ever create art, a capability closely associated with human creativity. Kurzweil claims to have used the neural net approach to create a computer program that composes poetry. The program, called "Ray Kurzweil's Cybernetic Poet," analyzes the rhythm and language style of existing poems and creates its own original work. It can imitate the rhythm and word structure of one author or merge the styles of several. It even has rules to prevent it from plagiarizing. Below is a poem, taken from Kurzweil's book *The Age of Spiritual Machines* (p. 166), that the program generated after analyzing poems by Randi and Kathryn Lynn:

Long Years Have Passed

Long years have passed.
I think of goodbye.
Locked tight in the night
I think of passion;
Drawn to for blue, the night
During the page
My shattered pieces of life
watching the joy
shattered pieces of love
My shattered pieces of love
gone stale. *

---

*Poem by Ray Kurzweil's Cybernetic Poet created by Ray Kurzweil. Reprinted by permission.

The cybernetic poet is just a basic application of the neural net approach. Some computer scientists believe that neural networks will be used extensively in the future to provide for consciousness in machines.

Another potential method of generating consciousness with computer technology is to use programs that simulate evolutionary processes. The advantage of such programs is that they will evolve as they gain new information. They will try a variety of solutions to a problem and make a note of the best ones for future use. Computer scientists who advocate this line of research hope that, just as human consciousness evolved in nature, machine consciousness will evolve with the proper programming.

But, how will programs, evolutionary or otherwise, enable computers to understand the information that they process? Marvin Minsky, a Nobel Prize winner and one of the founders of AI, asks this question. What does it mean to understand? Minsky says that understanding is the connecting of one idea to many others and the ability to interpret that idea in many ways. If someone simply memorizes the phrase "the neuron consists of a cell body, axon, and dendrites," does that mean he understands the structure of a neuron? No, he doesn't know that the cell body nourishes the cell or that the axon is a single fiber that fires electrical signals or that the dendrites are thin, branching tendrils that receive signals from other neurons. To understand a concept, one at a minimum must be able to see it in many ways and connect it to related concepts. If a computing machine could analyze information in multiple ways and explore alternative connections to larger bodies of structured information, then perhaps the machine could understand.

Some doubt that these three approaches are enough in themselves to build conscious machines. They believe another element is necessary: social interaction. As we saw in the previous chapter, many agree that interacting with others is critical for the development of a sense of self. Scientists at the Massachusetts Institute of Technology (MIT) tried to take this additional element into

COG. Shown with Rodney Brooks, the leader of the project, COG is built to learn from human interaction.
(© Peter Menzel. www.menzelphoto.com)

account in the building of their robot, COG. They built COG, a humanoid robot from the waist up, on the premise that human interaction is necessary for human intelligence. The robot plays with a Slinky, uses a saw, turns a crank, swings a pendulum, and plays the drums. It also can recognize human faces and imitate things that people do. COG's creators are hoping that the robot's humanoid form will make people deal with it as if it were human, thus giving COG training in relating to others. They want COG to have social opportunities for learning that might help it develop intelligent human abilities.

From this and other developments in the pursuit of Strong AI, Kurzweil concludes that the rise of conscious machines is on the horizon. As he puts it, human beings are patterns that can be reconstructed using computing technology. If social interaction in conjunction with one of the three programming approaches does allow us to build a conscious machine, that would imply that we, too, are machines—we must have been organic machines all along. The question now is this: how will we know whether the machine is conscious?

FURTHER READINGS

Many of Ray Kurzweil's ideas in this chapter, including the two thought experiments, were taken from three of his papers: "Live Forever—Uploading the Human Brain," "The Coming Merging of Mind and Machine," and "My Question for the Edge: Who Am I? What Am I?" All of them, as well as many other essays, are available on Kurzweil's Web site: http://www.KurzweilAI.net. To learn more about his views on computer intelligence, look for *The Age of Intelligent Machines* and *Are We Spiritual Machines? Ray Kurzweil vs. the Critics of Strong AI*. The poem by Kurzweil's program was taken from his *Age of Spiritual Machines*. The statistics on chess came from an article on Brainencyclopedia.com called "Chess." The information on the games between Deep Blue and Gary Kasparov was taken from IBM.com. The ideas of strong and weak AI were created by philosopher John Searle and are discussed in many of his books, including *The Mystery of Consciousness*. The leader of the CYC project is Douglas Lenat. To learn more about his work, read his 1995 paper "CYC: A Large-Scale Investment in Knowledge Infrastructure." The commonsense statements discussed were quoted from that paper (p. 33). To find out about the ideas of Marvin Minsky, read his book *The Society of Mind*. More information about COG and other robotics projects can be found on MIT's artificial intelligence Web site at www.ai.mit.edu.

7

TURING'S TEST OF CONSCIOUSNESS

In his famous and controversial 1950 essay "Computing Machinery and Intelligence," British mathematician Alan Turing (known for helping Britain break the German secret military code during WWII) proposes a method by which one can determine whether a machine is conscious. Originally called the "imitation game," the method has come to be known as the "Turing Test."

The test works as follows. The computer being tested is put in one room. A human responder is put in another room. Outside the two rooms, an interrogator sits in front of a monitor that he will use to communicate with the hidden person and computer. However, the interrogator does not know whether he is talking to a human being or a computer; he has to figure out which one it is. Turing says that if the hidden computer can convince the inter-

The Turing Test. The interrogator is shown here communicating with an unknown entity on the other side. He is trying to decide whether it is a person or a computer.

rogator (assuming he is of at least average intelligence) that it is a person, the computer is conscious.

A classic example of a conversation program that underwent an informal Turing Test was ELIZA, a virtual therapist, created by Joseph Weizenbaum in 1966. The story goes that Weizenbaum wrote the ELIZA program as a simple experiment. At one point he left his office, leaving the ELIZA simulation running on his computer. His secretary noticed this and walked over to it. When the programmer returned, he found his secretary engaged in conversation with ELIZA. She did not realize that ELIZA was not a conscious person and claimed that the program helped her change her life.

Clearly, the secretary would not make a good interrogator for the Turing Test. Had ELIZA been given a valid Turing Test, the interrogator would have asked questions that involved creativity or something so random that no programming could have an answer prepared for it. According to Turing, if any machine can

ELIZA: Hello. I am ELIZA. How can I help you?

User: Hello, I have a problem.

ELIZA: How do you do. Please state your problem.

User: I think I am a werewolf.

ELIZA: How long have you been a werewolf?

User: At least since the last full moon.

ELIZA: Please go on.

User: It all started when I noticed a curious amount of hair on my chest.

ELIZA: Why do you say your chest?

User: I never had hair there before. It disturbed me.

ELIZA: Can you elaborate on that?

User: Well, I felt as if my body was becoming something beastly.

ELIZA: Do you wish that your body was becoming something beastly?

User: No, I want to be human of course.

ELIZA: You are being a bit negative.

A conversation with ELIZA. ELIZA can only handle simple types of questions and statements from the user.

User: Please write me a sonnet about your lost love, Erico.

ELIZA: Can you elaborate on that?

User: Which of the Three Stooges, would you say, is most similar to the way you perceive yourself?

ELIZA: Oh, I say is most similar to the way I perceive myself.

Testing ELIZA. When put against the scrutiny of the Turing Test, ELIZA doesn't stand a chance.

User: Which of the Three Stooges, would you say, is most similar to the way you perceive yourself?
ALICE: The big one, I think.
User: Why do you like him?
ALICE: I have a lot of good experience with him.
User: When did you meet him?
ALICE: It was at time index 357241.

User: Will you go out with me?
ALICE: Have you dated any robots before?
User: No, but I am open-minded.
ALICE: I think you might be happier with a human. Umm. Do you mind if I tell other people you are open minded?
User: I think I am in love with you though.
ALICE: Oh. You believe that? Thanks I like you a lot too Elie.
User: So you will go out with me?
ALICE: Interesting deduction. What makes you say that I will go out with you?
User: I think so because you said you like me.
ALICE: It seems we have a lot in common, Elie.
User: That is why I think we would make a good couple.
ALICE: Yeah that's right. By "we" do you mean you and me?
User: Yes.
ALICE: I get it.
User: Will you marry me?
ALICE: Why don't you just download me?

A conversation with ALICE. ALICE can handle a wider range of questions than ELIZA.

provide convincing answers to such questions for an extended amount of time, it must be conscious.

A more modern Turing program is ALICE (Artificial Linguistic Internet Computer Entity). ALICE was one of four finalists in the

2004 Loebner contest for "most human computer," sponsored by the Cambridge Center for Behavioral Studies, Massachusetts, and won a bronze medal in both 2000 and 2001. The contest awards a bronze medal and $2,000 for the program that most convincingly simulates human conversation. The contest awards a silver medal and $25,000 to any computer that can pass the Turing Test by exchanging text messages. A gold medal and a $100,000 prize are awarded to a machine that can pass the Turing Test on an audiovisual basis—meaning that the machine would speak and *look like* a human being and would sit face-to-face with the interrogator. ALICE is programmed to be able to answer a lot more questions than ELIZA and won't be stumped as easily. However, it is still not good enough to pass the Turing Test. No machine has ever won a gold or silver medal at the Loebner competition.

The original Turing Test assesses the ability of a machine to interact with people through conversation. The Loebner competition takes the test a step further by offering a prize for a machine that also *looks* like a person. MIT's Cynthia Breazeal is the creator of Kismet, a robot that can simulate human facial expressions and thus seems to have emotions. The

User: Have you ever been in love?
ALICE: I love my bot master.
User: What is his name?
ALICE: His name is Rich.
User: Describe him.
ALICE: Objective case of He, used as a predicate complement with a linking verb. Also standing for Imperial Majesty.

User: Do you think I would get sick if I licked a poisonous toad?
ALICE: That's what I asked you.

User: Would it be inappropriate to begin shaving my back hair in an elevator?
ALICE: I don't care.

Testing ALICE. Though better than ELIZA, ALICE is far from being able to pass the Turing Test.

Kismet. Cynthia Breazeal is shown here playing with her robot creation. When Kismet detects the bright yellow dinosaur in her hand, its facial expression changes to one of excitement. (© Peter Menzel. www.menzelphoto.com)

robot head has cartoonlike features that can be repositioned to simulate nine different emotional states. For example, if Kismet's camera sensors ("eyes") detect a very colorful object (like a ball that someone puts in front of it), Kismet's facial expression may turn to the state of "happy" or "interest." If the person makes a fast arm motion in front of the sensors, Kismet's response might be "surprise" or "fear."

Though Kismet's facial expressions are not realistic, the concepts used to develop the robot could be applied to create one with facial expressions that are. In fact, Cynthia Breazeal and her students have been working on a new project: Leonardo, a more sophisticated robot with over seventy motors in its face, neck, and arms. In addition to showing expressions, this furry robot can respond to touch. Though not good enough to be mistaken for a conscious being, Leonardo may be the most expressive robot ever

Cynthia Breazeal and Leonardo. This furry robot is programmed to have natural-seeming social interactions with people. (Sam Ogden/Photo Researchers, Inc.)

built, and future robots will be even more sophisticated. Combined with the conversation ability of a successful Turing program, a machine that has convincing human facial expressions might be a step closer to being considered fully conscious.

Turing predicted that by the year 2000, machines being tested would be able to fool the interrogator for about five minutes. It is now past that date; the goal of five minutes has not yet been achieved. An astute interrogator would not be fooled for that long. However, conversation programs are steadily becoming more sophisticated; programming techniques are evolving and becoming more pristine. The time may come when, depending on the interrogator, a computer *will* be able to seem human for about that length of time. Eventually, Turing said, they will become so advanced that nobody will be able to tell them apart from human beings—they will be conscious. All we need now is the technology to accomplish this.

FURTHER READINGS

Alan Turing introduced his test in his famous paper "Computing Machinery and Intelligence," published in the periodical *Mind* in 1950. It is widely reprinted in other books and online. The test is discussed in the majority of books that address the question of machine consciousness. Ray Kurzweil's *Are We Spiritual Machines? Ray Kurzweil vs. the Critics of Strong AI* is one example. ELIZA was introduced in a paper by Joseph Weizenbaum called "ELIZA—A Computer Program for the Study of Natural Language Communication between Man and Machine." You can read the paper at many sites online and even talk to ELIZA. To talk to ALICE, go to http://www .alicebot.org. Try giving ELIZA and ALICE your own Turing Test. Information on the work of Cynthia Breazeal can be found in many books on robots or technology magazines, such as *Wired*. There is plenty online as well.

8

SUPREMACY OF THE MACHINES

In 1965, Intel corporation cofounder Gordon Moore observed the trends in the growth of technology and made a prediction, now called "Moore's Law," that computer processing power would double approximately every eighteen months—an estimate that has proven to be too low. Recently, processing power has been doubling every year. Moore also thought that the high rate of growth would last only until 1975, but actually it continues to this day. Ray Kurzweil extends Moore's Law even further in his "law of accelerating returns." He says that the rate at which technology is improving is also accelerating each year and predicts that, by the year 2020, circuit components will be only several atoms in width. Kurzweil also says that around this time a $1,000 personal computer will have the processing power of a human brain (about 20 million billion calculations per second).

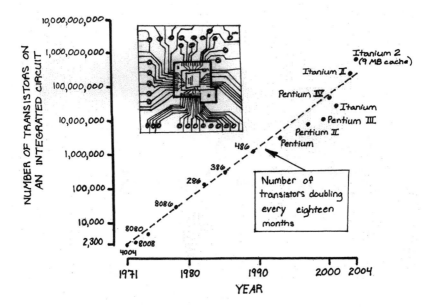

Moore's Law. The vertical axis represents the processing capability that will be available for computing (measured by the number of transistors in a circuit) that Moore predicted would double every eighteen months.

He says that, ten years later, $1,000 will buy a computer with the processing power of a village of human brains. By 2050, Kurzweil says, that amount will buy a machine with processing power equal to all the human brains in the world.

Through the use of "nanotechnology," the construction of objects atom by atom, we will be able to make computers that are even faster and much, much smaller. The technology has been used to build devices that can detect neuron firings and cause other neurons to fire. For example, when inserted into a living leech, the devices allow scientists to control the leech's movements with a computer. Even this feat is far from demonstrating the full potential of the processing power available today. Already, the processing power of your home computer is available in a chip that can rest on the tip of your finger.

Because of the recent surge in the power of computer technology, many believe not only that machines will one day be conscious, but also that they will have conscious abilities far greater than ours. This will be the age of superintelligence. Already, Kurzweil believes, machines have some degree of intelligence. He asserts that the reason people say the chess computer Deep Blue cannot think is that it can only perform calculations. Kasparov, he says, decides to make certain moves based on similar situations that he remembers from other games. Similarly, Deep Blue makes moves based on what is programmed in its memory. Therefore, Kurzweil says, Deep Blue engages in a level of thought similar to that of Kasparov. Kurzweil believes that if the programmers were to make Deep Blue's processing more similar to that of human thought, then the computer would be truly thinking. This will lead to progressively more advanced thought that will eventually dominate over human intelligence.

Marvin Minsky says that processes in computer chips already occur millions of times faster than those in brain cells. Therefore we should be able to design the superintelligent machines of the future to think millions of times faster than we do. Think of the efficiency that would result. Minsky says that, because of the incredible speed at which they will be thinking, an hour of existence might seem as long as an entire lifetime of a human being (just like the way I feel in math class) to these machines.

The road to superintelligence begins, as Kurzweil says, with uploading the mind onto computers. Once this is done, around 2030, we will be able to construct machines that will generate minds far superior to those of human beings. Kurzweil and Minsky agree that one of humanity's major setbacks is our unreliable organic makeup. We could each achieve so much more if it were not for all the failures of health that come with our slow, organic bodies. Unlike biological hardware, the parts used to build conscious machines will not decay so quickly. They will be more durable and last longer. They will also be easily replaceable.

With the proper assembly systems, we will be able to produce billions of replacement parts and copy whole machines quickly.

The machines will be able to share information instantly. We go to school, from preschool to graduate school, to learn what is already known about the world so that we may further that knowledge. This will be unnecessary in the age of machine intelligence because knowledge will be transferred from one being to another as easily as we transfer files between computers. Combined with the swift processing abilities of these beings, Kurzweil says, instant knowledge sharing will cause future technological progress to skyrocket.

Of course, we will also use this technology to enhance our existing organic machinery. Once our minds are stored in computer databases, we will have the option to be transferred into new, more reliable hardware. Nanotechnology will allow us to be replaced, atom by atom, with higher-quality materials.

Kurzweil and Minsky believe that this merging of man and machine is the next step in evolution on Earth. Natural evolution began very gradually, but later accelerated. The formation of the Earth took billions of years. Evolution began to occur faster with the appearance of multicellular organisms. The rate of evolution continued to increase until, at some point in time, a brain capable of conscious thought evolved. Similarly, technology evolved slowly. It took thousands of years for our ancestors to figure out how to use sharpened rocks as tools for survival. Today, the inventions of such things as the Internet have revolutionized the way we live in less than a decade. Next, Kurzweil and Minsky believe, evolution will bring man and machine together. We will live in a world where the line between human beings and machines is so blurred that people will stop making the distinction.

Kurzweil describes what this world might be like. It will begin in the middle of this century when intelligent machines begin to integrate into human society. Laws will begin to apply to them. It might be considered immoral to turn off a machine; it would be

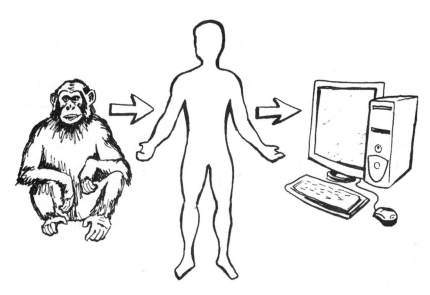

The next step in evolution. Kurzweil and Minsky predict that the next phase of human evolution will be the incorporation of synthetic technology into human bodies.

like killing a person. Such strange rules might arise because differences between machines and human beings will still be discernible, but that will change.

By the end of this century our organic bodies will cease to be important. Once we have been scanned and uploaded into computers, we will be able to switch bodies at will. Since mortality is linked to the decay of our bodies, it will no longer be a concern. We will be able to live forever. We will consistently create backup copies of our minds so that if there is an accident, and one's mind is destroyed, it can be easily reconstructed.

Internet connections in our brains will allow us to connect to the Web at any time, Kurzweil says. However, the Internet will be used very differently by 2099. Instead of merely visiting Web pages as we do today, we will enter virtual worlds that are as convincing as the real world. Instead of calling a friend on the phone, you will be able to meet him or her at a virtual Parisian café or at the top of a virtual mountain with a scenic view. This virtual life will be as detailed as real life in every way. We will be able to enter and exit instantly, and what happens to us in the virtual world will have no bearing on what happens to us in the real world.

If you are bored at home one day, you will be able to go hang out in virtual Rome or play basketball in a virtual gym. You could also enter a virtual home that you find more exciting. You could invent a new, virtual life with a new house and family, new friends, and a dog named Merbob. If someone is annoying you, you could escape into virtual reality. If a virtual person is annoying you, you can shut him or her off.

In the virtual world, Kurzweil says, people will be able to have any type of interaction with one another, including business meetings, social events, and even sex—regardless of their physical distance from one another. Yes, by the end of the century people will be simulating having sex without ever touching each other or even being in the same country. The end of the century will bring new meaning to the phrase "long-distance relationship." Sex will also be safer (unless there are virtual STDs, of course) and, according to Kurzweil, much better.

By 2099, human and machine intelligence will have almost completely merged. Most conscious beings won't have a body by this time. The human beings and machines that do have bodies will be indistinguishable from one another. Organic bodies will be a thing of the past.

Kurzweil and Minsky both predict that we are moving toward an age in which intelligent machines take our place as the dominant creatures on Earth. Since we are all machines, we can and must improve ourselves with better technology. They say that this is the next phase of evolution on our planet. The inevitable merging of man and machine is on its way. Machines more intelligent than we are will inherit the Earth.

FURTHER READINGS

To find out more about Ray Kurzweil and his views on AI, read his book *The Age of Spiritual Machines*. Also look for his papers "Live Forever—Uploading the Human Brain" and "The Coming Merging of Mind and Machine." Both papers are available on Kurzweil's Web site, www.KurzweilAI.net. Marvin Minsky writes about the future supremacy of the machines in his *Scientific American* article "Will Robots Inherit the Earth?" Also look for his book *The Society of Mind*. For many more essays on AI technology, visit Kurzweil's Web site. A good set of responses to Kurzweil can be found in his book *Are We Spiritual Machines?*

9

THE CHINESE ROOM

In 1980, philosopher John Searle devised what has become one of the most famous arguments against Strong AI (the belief that computer programs can generate consciousness): the Chinese Room. The argument takes the now-familiar form of a "thought experiment." The experiment works as follows: imagine the computer as a room. Instead of having circuitry, the room has inside it a man who can speak only English. On one side of the room, there is a slot through which the man receives written questions (input) in Chinese from someone outside the room. On the opposite side there is another slot through which he passes the answer (output), also written in Chinese. Inside the room, there is a rulebook (a program) that instructs the man as to what symbols he should return in response to the symbols he receives (the rulebook tells him how to process the information).

The Chinese Room. The person outside the room is shown here passing a message in Chinese into the room. The man inside then uses a rulebook that tells him what symbols to return. Though the man answers the questions appropriately, he doesn't understand a word of Chinese.

The man in the room provides satisfactory answers to the questions, but does he understand Chinese?

The answer is no. The man doesn't know a word of Chinese. He receives a question, follows a set of instructions, and passes out the answer—all without understanding a thing. When we communicate in English, we are not just mechanically moving random symbols around. We attach meaning to the words.

What is really going on here is that the Chinese Room is being given the Turing Test in Chinese. Say we put the Chinese Room in one walled area and a woman fluent in Chinese, Ms. Lee, in a nearby area. The interrogator writes questions in Chinese on a piece of paper and slips it through one slot for the Chinese Room and another slot for Ms. Lee. Let's say the first question is the Chinese translation of "What is morality?"

When the man in the Chinese Room receives the slip of paper with the question on it, what does he see? He sees Chinese symbols that to him are just meaningless squiggles. He shrugs and looks the symbols up in his rulebook. The man follows the instructions in the book and passes out another sequence of squiggles that, unbeknownst to him, means "concern with the distinction between good and evil or right and wrong; right or good conduct."

What about Ms. Lee, the woman fluent in Chinese? When she receives the slip of paper with the question on it, she does not see mere squiggles; she sees a deep philosophical question. The word "morality" strikes her deeply. She thinks about the relationship between good and evil in classic literature and in her own experience and considers the question through a lens of her own religious beliefs. The question not only has a simple meaning for Ms. Lee but also a profound impact on her consciousness. She thoroughly considers the many facets of the question and is troubled by having to provide a quick answer to a difficult problem. Nevertheless, she answers (in Chinese) that morality is "the relationship between the complementary forces of good and evil that exist within every human being."

Both Ms. Lee and the Chinese Room effectively pass the Turing Test. In fact, their answers are not only both convincing but also very similar. However, only Ms. Lee *understands* the question. The Chinese Room shows that it is possible for an entity to communicate as if it were conscious (as in the Turing Test) without actually understanding.

Searle uses the Chinese Room thought experiment to argue that the mind cannot be merely a computer program and the brain merely a computer. A computer follows a set of instructions (the program) and, like the man in the room, doesn't understand any of it. Thus the computer is not thinking. Searle says that his argument takes this logical form:

1. Computer programs work by manipulating symbols. (The man in the room sends and receives Chinese symbols.)
2. Minds have consciousness and allow for understanding. (In order to understand Chinese, one must have the mental content necessary to attach meaning to the symbols.)
3. Symbol manipulation alone is not enough to guarantee understanding or the presence of consciousness. (Though the man sends back the correct Chinese symbols, he does not attach any meaning to them, so he doesn't understand Chinese.)
4. Therefore, executing a computer program is insufficient for a mind to exist.

He says that as long as computers follow programmed instructions, they cannot be conscious. Just because something acts as though it is conscious, that doesn't mean it actually is conscious.

Searle rejects Kurzweil's claim that modifying the way the Deep Blue program works would give it true intelligence. In order to counter Kurzweil more directly, Searle presents a modified version of the argument, calling it the "Chess Room." Now imagine that, instead of receiving questions in Chinese, the man in the room, who has never played chess, receives a set of symbols, which he does not understand, that represent the positions of chess pieces on a board. The man looks up in a rulebook the move that should be next, and passes out the corresponding symbols. Although to people outside the room it would seem as if the man in the room understands chess, he doesn't even know that he is playing chess—he's just manipulating symbols according to a set of rules. He doesn't understand chess. He doesn't know Chinese. According to Searle, he doesn't know anything.

Those who believe that computers are, or can be, conscious are mistaking simulation for duplication, in Searle's view. He agrees that programs can *simulate* various aspects of consciousness. The programs ELIZA and ALICE, for example, can *simulate* the speaking and understanding of language, but they cannot *duplicate* those processes because they cannot create understanding. This is the same with the Deep Blue program. It only simulates an understanding of chess. There are also computers that can simulate fire and thunderstorms, Searle says, but they cannot duplicate them by setting rooms ablaze or striking office buildings with lightning. Similarly, computers can only simulate human consciousness.

Searle concludes that we are not speeding toward an age of conscious machines, and that no matter how powerful computer processors become, they will not provide consciousness for machines the way the brain provides it for us. Consciousness cannot be reduced to the calculation and symbol manipulation of computer circuitry. The sanctity of the human mind will be preserved.

FURTHER READINGS

The Chinese Room example and all its variations were created by John Searle. The thought experiment was first introduced in his 1980 essay "Minds, Brains and Programs," which is widely reprinted online. It is also discussed in his books *Minds, Brains and Science; The Mystery of Consciousness;* and *The Rediscovery of Mind.* If you are interested in what other philosophers have to say about Searle's thought experiment, try *Views into the Chinese Room* (edited by John Preston and Mark Bishop), a collection of essays by Searle, his critics, and his supporters.

10

DEMONS IN THE BRAIN

There are a number of philosophers who do not agree that the Chinese Room argument eliminates the possibility of programming conscious machines in our time. They claim that, though none of the individual elements in the Chinese Room (the man, the room, the rulebook) understand Chinese, the system as a whole does. This may sound strange at first. If the man doesn't understand Chinese, why should the man combined with paper instructions and wooden walls understand Chinese?

The brain is made up of different elements, none of which understands on its own. Yet, when the parts are put together, understanding takes place. Consciousness can be brought about only when all of the brain's components are combined and organized.

Philosopher David Chalmers proposes a thought experiment to illustrate this point. Suppose that one of the neurons in a Chinese speaker's head is replaced by the man in the Chinese Room, who for this experiment we will refer to as the *demon* (though I'm sure he's nice once you get to know him). The demon duplicates all the operations of the neuron it replaces and follows the same rules. When it receives a signal from neighboring neurons, the demon makes the necessary calculations and sends appropriate signals to other neurons.

So far, Chalmers says, the Chinese speaker's consciousness remains the same. The demon works exactly like a neuron and operates under the same set of rules. Now suppose that one by one, the other neurons in the Chinese speaker's brain are replaced by little demons until none of the original neurons are left. The demons have taken over all the signaling in the brain, but the Chinese speaker still maintains his consciousness. He has no idea that he has a skull full of tiny demons. He probably wouldn't believe it anyway. The network of demons works the same way as the former network of neurons.

Imagine now that we remove a single demon from the network. In its place, another demon doubles its workload. This demon will record the internal state of the eliminated demon on slips of paper,

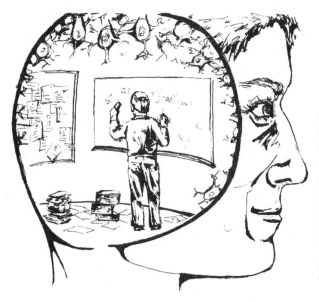

A demon in the brain. Chalmers says that this tiny person represents an entire network of neurons, just like the man in the Chinese Room.

updating and signaling when necessary, while also keeping track of its own internal state. We repeat this process by replacing other pairs of neighboring demons with individual demons that do the jobs of both. We continue to reduce the number of demons until eventually there is only one demon left, along with billions of slips of paper. These slips of paper are located in the positions of the original neurons. The lone demon scurries all over the brain, updating the billions of slips of paper (talk about writer's cramp) and accomplishing all of the functions of the original neurons.

This scenario is a copy of the Chinese Room. The demon, like the man in the Chinese Room, operates by manipulating symbols on slips of paper according to a set of rules. The demon and the man in the room are the same.

Chalmers says that his thought experiment shows how the system of the Chinese Room *can* account for consciousness. The Chinese speaker has the same consciousness with a single demon as he did with the original network of neurons. Just the same, the man in the Chinese Room represents the enormous network of parts needed to generate consciousness. The paper instructions are part of that network. Consciousness arises from the entire system, not from individual components.

From his technological perspective, Kurzweil agrees with Chalmers that the Chinese Room does not eliminate the possibility of programming conscious computers. He writes: "I understand English, but none of my neurons do." He claims that the man in the room represents the central processor of a computer, a small part of the system. The Chinese Room scenario leaves out all the other components of a computer over which the understanding is spread. After all, human understanding would be impossible without the brain's complex organization of interconnected neurons and chemical messengers. Just as there are many regions of the brain that interact, there are many parts of a computer that interact. No individual component of a brain or a computer has "understanding."

In his original paper introducing the Chinese Room, Searle listed a number of objections that he anticipated he would receive and wrote counterarguments for them in advance. Searle knew that someone would come along and say that the Chinese Room as a whole does understand Chinese even though the individual components do not. He called this objection the "Systems Reply."

To counter the Systems Reply, Searle points out that the rule-book and the room don't actually have to be there. The man could memorize the rulebook (then get rid of it) and work outside, but he still would not understand a word of Chinese. Even with the entire system contained within the man, there would be no understanding. Therefore, he says, the Chinese Room as a whole also doesn't understand.

Searle's opponents do not agree with this assessment. If he memorized all the rules in the book for how to get from one set of symbols to another, he *would* understand Chinese. He would be just like any fluent speaker of the language. Searle would insist, however, that, though any of us might be able to memorize a set of language symbols, we would still have no understanding of the language. Searle's opponents contend that the man memorizing the rulebook would be equivalent to our memorizing words in English. Nothing would be missing from the man's education. The interaction between the materials in the Chinese Room can produce consciousness just as the brain does.

David Chalmers supports this view. Conscious understanding, Chalmers agrees, is spread over a "functional organization" of matter. It can arise from a functional organization of any material. Chalmers defines functional organization simply as an arrangement of components that interact to produce some behavior. He says: "Whether the organization is realized in silicon chips, in the population of China, or in beer cans and ping-pong balls does not matter. As long as the functional organization is right, conscious experience will be determined."

To combat this concept, Searle refers to an argument called

the "Chinese Nation" (they seem to love incorporating the Chinese into every discussion). Imagine that every citizen of China is a functional unit in a system—they each represent a neuron, for example. If an interacting organization of components was enough to provide for consciousness, that would mean that, as a whole, the country of China could be a conscious entity—an intentionally absurd idea.

Chalmers, however, does not believe that just any functional organization has consciousness. A random network of neurons or demons randomly releasing signals would not constitute a mind. Similarly, the population of China is not properly organized to produce consciousness. We don't know how the components of the brain work together to create a mind, but somehow they are organized in the proper way. If the population of China were also organized in the proper way, it, too, would produce consciousness.

This may be hard to accept. How could any organization of the Chinese population produce consciousness? The notion seems preposterous. Nonetheless, it seems just as preposterous that a lumpy blob of gray tissue can produce consciousness. Consciousness can arise from a physical organization of parts regardless of what the parts are made out of, Chalmers says. The physical organization provides the basic mechanisms *necessary* for consciousness, but does not actually generate it on its own. He says that consciousness is something separate, as we will see.

FURTHER READINGS

Searle's discussion of the "Systems Reply" can be found in his 1980 paper "Minds, Brains and Programs." Again, it is widely reprinted online. Discussions of this and other replies can be found in *Views into the Chinese Room*, edited by John Preston and Mark Bishop. The demon thought experiment of David Chalmers is discussed in his book *The Conscious Mind*. His quote that I use was taken from page 249 of that book, as was Chalmers's definition of functional organization. Kurzweil's objection can be found in his book *Are We Spiritual Machines?* The Chinese Nation argument was originally put forward by philosopher Ned Block. For an exchange between Searle and Chalmers, see Searle's *Mystery of Consciousness*.

11

DESCRIBING
THE INDESCRIBABLE

Imagine that you are at the beach. The August sunlight is beating on your back and the air is fresh. As you walk along the shore, you feel the sand between your toes and the mist from the ocean collecting in small droplets on your face. You taste the salty water as the drops drain into your lips. Every so often, a breeze ruffles your shirt and sweeps past you.

The water shimmers in the sunlight. You hear the roll of waves on the shore, the squawking of seagulls overhead, the distant rumble of the highway you took to get here. You pass hundreds of people as you walk. There are kids playing volleyball to your right. You hear them laughing. Suddenly, from the corner of your eye, you see a colored object flying at you. You duck just in time. A little boy runs toward you, apologizes, and picks up his beach ball.

As the sun sets, you see the brilliant red-orange of the sky reflected on the water. You sit on the shore and watch.

All of these events are parts of one day at the beach. All of your five senses are combined to create one unique and richly detailed experience. No one else at the beach has the same experience as you. Your experiences are private and nobody will ever know exactly what they are like.

These experiences, which philosophers call *qualia*, are extremely difficult to describe. Could you describe the taste of salt to someone who has never tasted it? Could you explain how it feels to see a sunset? What about the sound of ocean waves? Let's say you tried as best you could to describe to someone how ocean waves sound. Would they know how it feels to hear them?

Qualia (the singular form is "quale") are the special, indescribable, inner experiences we all have. The experiences of feeling sand between your toes, hearing the sound of rolling waves, and watching a sunset are examples of qualia.

Qualia play a very important role in the debate about consciousness. David Chalmers considers qualia to be at the core of consciousness. He says that "to have qualia" means almost the same thing as "to be conscious." We each know that we are conscious because we have qualia. A robot might claim to be conscious and say that it feels the sand between the pressure sensors on its metal toes, but if qualia are absent, it is *not* conscious.

Like the Chinese Room, the concept of qualia can be used to evaluate behavioral tests of consciousness. Recall that the man in the Chinese Room can answer questions well enough to pass the Turing Test, but the test appears to be invalid because he doesn't *understand*. Similarly, if an entity *behaves* as if it is conscious, but has no qualia, it is not conscious.

The concept of qualia can be put into perspective by a famous question in the philosophy of mind: What is it like to be a bat? We know bats are mammals with the ability to fly. We also know that bats are blind. In order to get around obstacles, they use a sonar system called *echolocation,* which works as follows. The bat emits a series of high-pitched shrieks that create echoes when they reflect off any object in range. These echoes are received by the bat and analyzed in its brain. From the echoes' quality and the time it takes them to return, the bat can judge its distance from an object as well as that object's size, shape, and texture. It perceives the world in a way that is completely different from the way we do.

With all the scientific knowledge we have about bats, do we know what it is like to be a bat? Do we know what it is like to have a bat's experience? A bat's perspective? The answer is no. Knowing the physical and functional characteristics of a bat is not enough to tell us what it is like to be one. Though we know how echolocation works, we don't know *what it is like* for a bat to "see" using a sonar system. We don't know what it is like to have a bat's qualia or what it is like to be a bat from the inside. Nevertheless, novelist Margaret Atwood tries to describe it in her short story "My Life as a Bat." Here is an excerpt:

> . . . I am clinging to the ceiling of a summer cottage while a red-faced man in white shorts and a white V-necked T-shirt jumps up and down, hitting at me with a tennis racquet. . . . A woman is shrieking, "My hair! My hair!" and someone else is calling, "Anthea! Bring the stepladder!" All I want is to get out through the hole in the screen but that will take some concentration and it's hard in this din of voices, they interfere with my sonar. There is a smell of dirty bathmats—it's his breath, the breath that comes out from every pore, the breath of the monster. I will be lucky to get out of this alive.
>
> . . . I am winging my way—flittering, I suppose you'd call it— through the clean washed demilight before dawn. This is a desert. The yuccas are in bloom and I have been gorging myself on their juices and pollen. I'm heading to my home, to my home cave, where it will be cool during the burnout of day and there will be the sound of water trickling through limestone, coating the rock

with a glistening hush, with the moistness of new mushrooms, and the other bats will chirp and rustle and doze until night unfurls again and makes the hot sky tender for us.

Though her description is wonderful, Margaret Atwood does not know what it is like to be a bat. In fact, the account is flawed. The bat would not be able to notice the color of the man's face or clothing. Bats are blind. Her story only tells us what she imagines it is like for her character to be in a bat's body. I could sew fabric wings to my shirt, tape my eyes shut, tie my feet to the ceiling, and eat a handful of flies, but that would miss the point. The experience would only tell me what it is like for me to pretend to be a bat.

I would find the taste of flies revolting, but a bat probably finds them delicious. It might fly over to its buddy and squeal about how the large ones are especially juicy. I would not experience the same *qualia* as a bat because I am not a bat. Margaret Atwood isn't either. We lack bat qualia. Neither of us knows what it is like for a *bat* to be a bat.

The thought experiment of the bat shows two things: that conscious-seeming behavior is not enough to demonstrate that one has qualia, and that the qualia of an entity are separate and cannot be determined from the physical facts about it. You can use this thought experiment to clarify which beings you believe are conscious. For example, if you think dogs are conscious, then you believe that dogs have qualia. You believe dogs have a point of view—that the answer to the question "What is it like to be a dog?" isn't "Nothing." Do flies have qualia? What about flowers? Rocks? These are questions that you must answer for yourself because from the outside we cannot tell. Qualia are separate from observable data.

This is why many philosophers believe in dualism, the idea that consciousness is separate from the physical world. Since, they claim, there are conscious properties of things (qualia) that are beyond the physical properties, consciousness must be nonphysical.

Philosopher Frank Jackson uses another thought experiment to illustrate this point. Imagine that we live in the far future when our knowledge of neuroscience is complete and we know everything about how the brain interprets the outside world. Mary is a neuroscientist who specializes in color vision. For her entire life, she has been imprisoned in a black and white room. She has never seen any colors except black, white, and shades of gray (even Mary herself is painted these colors). Using black and white textbooks and a black and white television, she learns everything there is to know about color. Mary learns, for example, which colors refer to which objects. She learns everything about how light waves travel and how they are absorbed and reflected by objects. She is an expert on the structure and function of the eye: how it receives light waves, focuses them on the retina, and converts them into nerve signals that travel to the brain. She also knows how the photoreceptors in the eyes, called *cones*, allow for color detection. Finally, Mary knows everything there is to know about how the brain interprets the nerve signals it receives from the eyes and causes us to see color.

What happens when Mary is released from her room and sees a red apple for the first time? Will she say that the color red is exactly what she thought it would be or will Mary learn something new?

Jackson asserts that Mary *will* learn something. She would marvel at the depth and beauty of color. Though Mary masters the science of color vision, she still does not know what it is like to see color. There is something about the experience of seeing color that no amount of instruction can teach her. For Jackson, this demonstrates that having qualia (and therefore consciousness) is separate from physicality or behavior.

To connect the thought experiments, let's recall the robot Kismet that alternates between nine different facial expressions that make it seem to express feelings. Let's join Kismet for a day at the beach. There she lies on her towel (some refer to Kismet as "she"), under the sun, watching the waves roll in. The salt spray collects in her bushy eyebrows and drains toward her rubbery red lips. Kismet's ears perk up at the sound of seagulls and her eyes follow the beach ball rolling by, but are Kismet's facial expressions enough to guarantee that she is experiencing anything? No, her behavior does not demonstrate that she has qualia. There is nothing *it is like* to be Kismet. She has no point of view. Again, we see that, since qualia are essential to consciousness, a being must do more than act or seem conscious in order to be so. Without qualia, Kismet is going to miss a great sunset.

FURTHER READINGS

The question of whether we know what it is like to be a bat was introduced by philosopher Thomas Nagel in his 1974 essay "What Is It Like to Be a Bat?" It was originally published in the journal *Philosophical Review*, but it can also be found in many books such as *Philosophy of Mind*, a compilation of essays edited by David Chalmers. It is also available online. The thought experiment of Mary the color scientist was introduced in a 1982 article of Frank Jackson's called "Epiphenomenal Qualia." It was originally published in *Philosophical Quarterly* and is also widely reprinted. Margaret Atwood's short story "My Life as a Bat" can be found in many short story collections. I found it in the Atwood collection *Good Bones and Simple Murders*, pp. 109–16. The story is not geared toward Nagel's concept of the bat but is interesting nonetheless. More discussions of qualia can be found in *The Conscious Mind* by David Chalmers.

12

MARCH OF THE ZOMBIES

Imagine that you have an identical twin who shares not only your appearance, but your personality and behavior as well. Your friends are unable to tell the difference between you and your twin. Your parents wouldn't notice any difference if your twin showed up to Friday night dinner instead of you. The two of you are identical in every way except one: your twin is not conscious.

Your twin would be what philosophers call a *zombie*, a non-conscious entity that looks and behaves as if it were conscious. It is not some undead, slime-covered, skeletal creature come from the grave to feed upon the living. The philosopher's zombie has no relation to the classic Hollywood zombie. Philosopher David Chalmers writes: "A zombie is just something physically identical to me, but which has no conscious experience—all is dark inside." Zombies act as though they are conscious but lack qualia

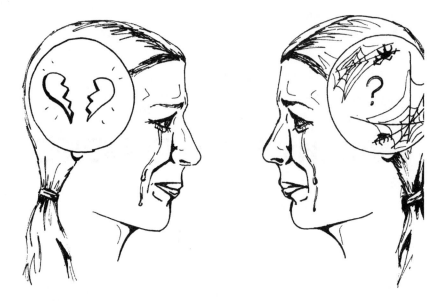

Zombie Twin. From the outside, one would not be able to tell the difference between a zombie and a person, but the zombie is not conscious.

and the capability to make decisions the way that humans do. For example, suppose that your zombie twin picks a rose and says: "Wow, look at the rose I picked! It is such a deep red and it smells wonderful." The zombie seems to *decide* to pick the rose and to *enjoy* the smell and color, but actually it does none of these things. From the outside, everyone would assume that your twin is conscious, because, as we know, qualia are private, inner experiences. There is no "mark of zombiehood" that would reveal whether an entity is a zombie. Whenever you and your zombie twin engage in the same activity, the two of you will act alike, but only you will have qualia. The zombie knows nothing and experiences nothing. There is nothing it is like to be a zombie.

How could a zombie come about? Perhaps one could be built out of silicon chips and programmed to behave like a human being, or maybe it could be made out of biological parts. It doesn't actually matter how a zombie is created as long as it looks and behaves as if it were conscious.

Many people think that Chalmers is simply saying that, if zombies walked the Earth, there would be no way to tell—based on their behavior—that they aren't conscious. However, he is actually saying much more than that. Chalmers argues that, since the zombie can do whatever a person can do and do it without being conscious, *consciousness must not be an essential part of our behavior*. In other words, each of us might behave the same way without our consciousness. Consciousness is merely an "optional extra." Adding a homunculus (the little man inside your head who represents your consciousness) to a body adds qualia, not function.

From here we see why David Chalmers believes in *functionalism*. Functionalism is a materialist belief that mental states can be explained as the interactions between the physical components of the brain (or any other organization of parts, such as a computer). So, rather than being an organization of physical parts, a thought is a pattern of information flow between those parts. According to functionalism, a person could still behave normally without consciousness because the pattern of interactions in his or her brain would still work. Only qualia would be lost.

Imagine a pilot in an airplane. The pilot can leave the cockpit and the plane will fly just as well on autopilot. Without him, there is nobody to make the *conscious decisions* to steer the plane or lower its altitude; those functions will take place automatically. A little girl watching the plane from the ground cannot tell that the pilot is not in the cockpit. The plane is flying just like any other plane. Surely someone must be flying it. The girl doesn't know that it is a zombie plane. She cannot tell that there is no pilot inside to marvel at the beauty of the sky.

One might be wondering how a zombie could exhibit human behavior without being conscious. Isn't consciousness necessary to control our behavior? Though no program has been able to pass the Turing Test so far, they may be able to do so in the future. Recall that the Loebner competition awards a gold medal for a machine that can pass the test on an audiovisual basis, meaning that the machine would have to convince the interrogator of its humanity by acting and appearing conscious. This is extremely difficult, but certainly possible. A machine that could impeccably pass this test would be a zombie. It is possible that, years from now, engineers will develop such machines. Chalmers says that if these machines were released into the general public, they would be treated just like regular citizens because nobody would be able to tell that they aren't conscious. After a while, many would accept the zombies as human; zombies would integrate into society (I see a reality TV show coming). Nobody would know the difference.

Since it is reasonable to suppose that a nonconscious entity can have the same physical organization and behavior as a human being, it is logical to conclude, Chalmers says, that consciousness does not result simply from behavior or physical components. In other words, since the zombie can have human structure and behavior without having consciousness, behavior and structure do not create consciousness. The intricate network of neurons in the brain does not create consciousness by itself; consciousness is more than a mere physical phenomenon.

Now imagine a zombie world where the natural rules of our world—those that cause consciousness to arise from a suitable arrangement of parts—do not hold. This is a world just like ours, except that everyone is a zombie. Let's call it Zombie Earth. Every person in our world has a zombie twin on Zombie Earth. Aliens visiting Earth and Zombie Earth would not be able to tell which of the worlds they were visiting because the two worlds are physically identical. The only difference is that Zombie Earth is a world without consciousness.

Since it is logically possible, Chalmers believes, that the zombie world is particle-for-particle the same as our world, but lacks consciousness, it follows that consciousness is something additional to and separate from the physical world. The same con-

Zombie Earth. This is a world physically identical to ours, but lacking in consciousness.

clusion can be drawn from a similar thought experiment. Instead of Zombie Earth, where there is no consciousness, imagine Inverted Earth, where all conscious experience is inverted. For example, while we have blue experiences when looking at the sky, inverted earthlings have red experiences. Inverted earthlings have hot experiences when they jump into freezing water and cold experiences while tanning at the beach. Though the inhabitants of Inverted Earth are physically and behaviorally identical to us (and zombies), their qualia are different. The Zombie Earth and Inverted Earth thought experiments reinforce Chalmers's conclusion from the individual zombie: consciousness is an aspect of the world that transcends the physical facts about it. Consciousness cannot be a creation of our bodies because it is logically possible for our precise behavior and physical structure to remain the same while consciousness is either absent (zombies, Zombie Earth) or altered (Inverted Earth).

Rather than believing that an organization of physical parts creates consciousness, Chalmers says that the organization *gives rise* to, or provides the foundation for, consciousness. The distinction is subtle and may be hard to grasp. Think about electricity. Though the potential for electricity exists everywhere, we cannot *create* it. We can only harness it. When we turn on a light, we complete a circuit. We create the proper organization of parts to harness electricity. However, that doesn't mean that the circuitry in a light can *create* electricity; energy cannot be created or destroyed. When the circuit—the organization of parts—is complete, electricity will flow through the wires and the bulb will light up. Similarly, according to Chalmers, a physical arrangement of parts (such as the brain) does not create consciousness; it allows consciousness to arise.

How might consciousness arise? Chalmers says that just as there are fundamental laws of physics, there are fundamental laws of consciousness. These laws dictate the conditions necessary for consciousness to arise from physical components. We do not yet know what these are, but according to Chalmers they exist, just as the laws of physics existed before Newton's discoveries. These laws would state how physical components are related to mental components. They would show how consciousness relates to physical organization while still being separate from physical things.

But wait—this view is strongly dualist. Recall that dualism is the belief that consciousness belongs to a category distinct from physical things. This is the view that Chalmers seems to support here. However, we saw earlier that he believes in functionalism, the materialist idea that consciousness is nothing more than the interactions between physical parts. These views seem to contradict each other. How does Chalmers resolve this?

As a functionalist, Chalmers believes that mental states are just the interactions between properly organized physical components. This is why the existence of zombies is a logical possibility.

However, unlike other functionalists, Chalmers does not believe that consciousness can be simplified to such interactions. Consciousness, he says, is something *irreducible*, something that cannot be explained simply in physical terms. Functionalism can go only as far as to explain the physical events in our body that take place at the same time as the conscious events.

Imagine that you slap your zombie twin across the face (apparently you didn't like the idea of someone copying your functional organization). Functionalism explains all the physical events that occur, such as neurons in the zombie's body shooting signals to its brain, which registers the pressure that you put on your zombie twin's face. Functionalism would also explain why the zombie might run away (to avoid receiving another blow) or slap you back in retaliation. These are processes that can be explained, Chalmers says, merely by the organization of physical parts. Consciousness is separate from these physical interactions.

To accommodate consciousness, Chalmers promotes a specific type of dualism called "property dualism." According to this belief, instead of separate physical and mental worlds, there is one world where everything has both physical properties and mental properties.

So, let's say that your zombie twin continues the sibling rivalry and does slap you back. You, unlike your twin, will consciously experience pain. However, Chalmers says that there are really two meanings to the word *pain*. The first is the unconscious functionalist meaning, according to which pain is a pattern of neuronal activity. This occurs in both you and your zombie twin. The second is the conscious feeling of unpleasantness that only you experience. From here we see that the conscious aspect of pain *mirrors* the physical, or functional, aspect. Changes in functional organization do not cause changes in consciousness, according to Chalmers. They just occur with perfect, parallel correlation. The difference in the body's physical structure (different patterns of neuron signaling, etc.) during pain is mirrored by the

Fading qualia and dancing qualia. Chalmers contends that these are impossibilities.

qualia of suffering from pain (a change in consciousness). The physical events that occur when we see a field of grass are accompanied by the qualia of seeing green. As someone's brain processes the smell of a rose, he or she will consciously experience its pleasant scent. Every change in physical organization is mirrored in its conscious counterpart.

As you go to sleep at night, your physical body begins to shut down. At the same time, your consciousness begins to fade. When someone is injured and suffers brain damage, his or her consciousness will be altered.

Chalmers asserts that it is impossible for there to be a change in physical organization without a corresponding change in consciousness. It would be absurd, he says, for qualia to fade without the body's systems also shutting down. He calls this impossible scenario "fading qualia." He says it would be equally absurd to have "dancing qualia," when one's qualia suddenly become much stronger without a corresponding change in functional organization. Changes in the physical and the mental will always coincide.

Now, how do all these ideas of zombies, zombie worlds, inverted worlds, functionalism, and property dualism work together as a theory of consciousness? Think about your day at the beach from the previous chapter. The theory works as follows. All the experiences you have at the beach occur in two ways: they cause changes in your physical body and corresponding changes to your consciousness. We know, Chalmers says, that physical events and conscious events are separate because it would be logically possible to have a world—with natural laws different from those on Earth—in which the same physical events occur *without* conscious events (Zombie Earth) or with *different* conscious events (Inverted Earth). Though they are separate, physical events and conscious events will always occur together, such as when you see a sunset. As you see a sunset, the physical event consists of the frequencies of light striking the receptors in your eyes and being transformed into electrical signals that travel along paths of neurons to the brain, which processes the signals as an image. The conscious event—which occurs at the same time—is the experience of seeing and appreciating the beauty of the color. Functionalism explains the physical event whereas property dualism explains the physical as well as the simultaneous conscious event. The zombie has only the physical events (enough to exhibit human behavior). We experience both physical events and conscious events. This is the theory of David Chalmers.

As all theories do, this theory has its critics. Since Chalmers's theory rests mostly on the conceivability of zombies, many opponents attack this idea first. One critic, Todd Moody, tries to show that a zombie *cannot* exhibit the same behavior as a human without having consciousness. Moody asks us to imagine Zombie Earth once more.

On Zombie Earth, everything is supposed to seem the same. Zombies go to high school and college and get jobs in corporations, just as we do. Zombies go on dates and get married and have baby zomblings together. There are annoying zombie neighbors and old, heavyset zombie aunts, zombie hair stylists, and overpriced zombie dentists. Of course, there would also be zombie philosophers. Some of these philosophers would study the philosophy of mind, just as philosophers do in our world. However, Moody says that zombie philosophers would be awfully "confused" by the notion of consciousness. They might never even come across the concept. Why would they? Therefore, zombie philosophers wouldn't come to the same conclusions as conscious philosophers. There would be many other differences too, not just in philosophy. So, it appears as if there *would* be differences between Earth and Zombie Earth. The zombie thought experiment doesn't work. Moody says that consciousness is not simply an "optional extra"; it is required for many of the things that human beings do. According to Moody, the very idea of a zombie is inconceivable.

Chalmers, however, stands by his theory, maintaining that there would be no observable differences between our world and the zombie world. If you accept Chalmers's theory, then you believe that human beings cannot be machines. Machines, just like zombies, do not have conscious properties. Chalmers maintains that what sets us apart from the zombies is that the physical organization of our brains connects us to the realm of consciousness, providing us with an irreducible experience that machines will never have.

FURTHER READINGS

The chapter focused mainly on the work of philosopher David Chalmers. Most of the ideas were taken from his book *The Conscious Mind*. Chalmers's quote at the beginning of this chapter came from page 96 of the book. Other concepts were taken from an essay by Chalmers titled "Consciousness and Its Place in Nature." It can be found in his edited volume *Philosophy of Mind*. For more information on Todd Moody's attack on the zombie concept, look for Moody's essay "Conversations with Zombies." It is available online. An exchange between Searle and Chalmers can be found in Searle's *Mystery of Consciousness*.

13

THE DENIAL OF CONSCIOUSNESS

Philosopher and cognitive scientist Daniel Dennett asks us to imagine a special kind of zombie—one that can monitor itself. This zombie's computing system processes information at several levels. Imagine that it has processing levels from A to Z, level A being most complex and Z being least. Processing level A monitors level B, which, in turn, monitors level C and so on. As a result, this zombie reviews its own internal activities. Let's call this zombie a *zimbo*. A zimbo is just a complex zombie that has self-monitoring capabilities.

Now imagine that you give a zimbo the Turing Test. You ask it how many hours are in a week. It responds by saying that there are 168 hours in a week. You then ask the zimbo how it came to that conclusion. Since the zimbo is self-monitoring, it can review its own processing, analyze it, and say that it took the number of hours in a day (which was stored in its memory) and multiplied it by the number of days in a week (also in its memory). The zimbo appears to "know" how it found the answer.

Dennett is convinced that the zimbo would be able to pass the Turing Test. Here we have a nonconscious machine capable of self-monitoring and seemingly intelligent behavior.

When the zimbo reviews its processing states and speaks about them, it would unconsciously believe that it was in those states. In other words, *"It* would think it was conscious, even if it wasn't!" Dennett says that any such machine that could pass the Turing Test would be the victim of an illusion caused by its own machinery—the illusion that it was conscious.

Dennett believes that we human beings suffer from the same illusion. We are nothing more than complex machines with many processing levels, the higher levels reviewing the lower ones. *Nobody is conscious,* at least not in the way we normally think of consciousness. There is no "ghost in the machine." There is no driver, no pilot. The machine operates itself. What we call "thinking" is just complex information processing in the brain (which works like a computer). Like the rising and setting of the sun, consciousness is an illusion.

In order to understand what Dennett thinks we are, it helps to understand what a parallel computer is. A typical home computer has only one central processor that controls its operations. A computer that uses parallel processing (a parallel computer) spreads its operations over several processors. The combined power of the processors allows many programs to run at the same time.

Think of the brain as a parallel computer. Mental activity is carried out by parallel processing. Information (input) is accepted and interpreted by a continuous process that Dennett calls "editorial revision." Let's say you are sitting in the back of a car, looking out the window. As the car moves, your eyes continuously receive new information in the form of reflected light waves and transmit it in the form of nerve impulses to the brain.

The parallel processing machinery constantly reviews and edits the incoming data in order to update the image that you see. Dennett says that these editorial processes take only fractions of a second. What you interpret as an image in your mind is actually the result of many revisions (additions, subtractions, formattings) made to the visual signals by complex, parallel circuitry.

Dennett also says that when we "learn" we simply acquire new programming. When we come upon new ideas, catch phrases, songs, or fashions, our programming expands to include them. When we succeed or fail at doing something, we store that information in our memory, allowing us to use it when we make decisions in the future. Just as the Deep Blue chess computer decides which move to make by referring back to lists of past games (stored in its memory), we choose what to do based on our past successes or failures and the routines that our higher-level programs create as we go through life.

Dennett calls his theory the "Multiple Drafts" model of consciousness because during processing (editorial revision), the brain creates many drafts, or interpreted versions, of the incoming data using its parallel machinery. He says that it will ultimately replace Descartes's old theater model of the mind.

As part of his theory, Dennett denies that there is good reason to believe that qualia exist. He says that when we experience what we think are qualia, we are actually only processing data. We think of our regular brain processing as indescribable, inner experience, but this is an illusion caused by our machinery.

When philosophers talk about qualia, they most often use color as an example. What is color? Each color is associated with a certain frequency of visible light. Colored objects have certain reflective properties that cause them to absorb some frequencies and reflect others. What is happening, according to Dennett, when we see color?

When you look at an apple, it is reflecting specific frequencies (wavelengths) of light. Receptors in your eyes, called *cones*, receive the light frequencies, convert them to nerve impulses, and send them to the brain. The brain accepts these signals as sensory input (data). Next, the brain processes the frequency along its parallel circuitry. What we call the experience—or qualia—of seeing the red color is the final interpretation of the light frequency. There is nothing "special" about this process. It is completely mechanical.

Robots see (or rather, process) colors in the same way. If a robot with visual sensors is programmed to figure out the difference between two colors, it would process the data just as we do, Dennett says. Let's say that an image of Santa Claus and one of the American flag are placed side by side, and a robot must figure out which is darker: the red on Santa's suit or the red stripes on the flag. The robot would focus on each one and, with its visual sensors, receive the reflected light as input. Next it would process the information, assigning values to each color: red #163 to the stripes on the flag and red #172 to the Santa Claus suit. It would then compare the values by subtracting 163 from 172 and getting 9. The robot's conclusion would be that the Santa Claus suit is a slightly darker shade of red. When we see color, we use another sort of mechanical process. Electrical signals and neuron firings along our brain circuitry are what allow us to interpret color. There are no qualia.

But what about the thought experiment of Mary, the color scientist? Surely that demonstrates that qualia exist. When Mary leaves the black and white room, wouldn't she learn something new? If she were to see a rose or an apple, wouldn't she be surprised by the way the color red looks?

Dennett replies that she would not be surprised. If she learns *everything* there is to know about color and light waves in her black and white room, then she would also know the effect that seeing color would have on her nervous system. She would know how the various light frequencies would make her feel and what thoughts she would have upon seeing them.

He writes a new ending to the story. One day, Mary's captors decide to release her. Attempting to trick her, they hand her a blue banana immediately as she walks out of the room and into the world of colors. Mary looks at it and frowns.

"Hey, you are trying to trick me," she says. "Bananas are supposed to be yellow. This one is blue! I'm not going to fall for that."

Shocked, Mary's captors ask her how she knew.

"Well, when I was inside the room, I learned *absolutely everything* there is to know about frequencies of light, reflective properties of matter, and all the causes and effects of color vision. I learned exactly how each color would affect me—including the yellow of banana. Nice try though, guys."

In the room, Mary learns how the parallel architecture of her brain would interpret all frequencies of light in order to construct an image of her surroundings. There are no "qualia" that she was missing. The physical information was all Mary needed. In fact, according to Dennett, that's all the information there was. When Mary leaves the room, she learns nothing.

The Multiple Drafts model also accounts for changes in our senses. Think about this: coffee is said to be an acquired taste. Coffee drinkers often say that when they first tried it they hated the taste. Now, they adore that same bitter taste and drink coffee every day. What has changed? The coffee itself hasn't changed—the beans are grown and prepared in the same way. According to the theory, what has changed is the set of programs through which the coffee drinkers interpret the data sent by nerves from the taste receptors. Just as they incorporate songs and fashion into their programming, they can acquire routines that make them "like" the taste of coffee.

Let's say that Jerome, twelve years old, is not a coffee drinker, but he notices that coffee is a big part of our society. He notices that many people seem to enjoy drinking it. He tries it occasionally, noting the taste. These observations are stored in his memory. As he gets older, these incidents lead him to develop new processing routines. The operation of his circuitry evolves to a point at which the higher-order programs in his brain interpret the data in a new way. Now seventeen years old, Jerome has a cup of coffee and, to his surprise, likes it! The way his brain interprets the chemical signals sent by nerves in the tongue has evolved. Jerome's changed programming has made him a coffee drinker. Despite what he thinks he is sensing, qualia have nothing to do with it. He is no different than a complex, self-monitoring zombie: a zimbo.

Dennett claims that his theory is consistent with science. In science, in order to be sure of something, one has to confirm it. The only things that exist in the world, he says, are those whose existence we can prove scientifically. Consciousness is one of the few things that we *think* exists in the world but cannot prove the existence of. If I pinch my arm, only I will feel the resulting pain. Nobody around me will know that I feel it. I could *say* that I feel pain. I could yell "ouch!" but that wouldn't prove anything; I could be pretending. The qualia of tasting coffee, seeing a red apple, hearing a symphony, and feeling the pain from a pinch are all scientifically unconfirmable. They are illusions—they are the way in which we view our own data processing. To claim that we have consciousness is to diverge from science.

If we take Dennett's view, then the answer to the question posed by the title of this book is yes. Part of the reason that many are reluctant to accept these ideas may be that they are afraid to do so. As the use of man-made machines (robots) increases—as machines begin to take our place in various tasks throughout the world—we would like to think that we have something that they do not: consciousness. It is comforting to believe that no technology could ever duplicate the special and wonderful power that the human brain provides. We fear to accept the possibility that we are not special. Nevertheless, according to Dennett, this is the reality: there is no difference between a person and a zombie. Consciousness is an illusion. We are zombies.

FURTHER READINGS

The concepts in this chapter were introduced in an article by Daniel Dennett called "Quining Qualia." It can be found in *Philosophy of Mind*, edited by David Chalmers. Dennett's opinions are discussed more thoroughly in his book *Consciousness Explained*, published in 1991. The quote of his that I use was taken from page 311 of that book. The example of a machine distinguishing between the red on Santa's suit and the red on the American flag was taken from there as well. The comparison between a sunset and the illusion of consciousness, along with an interesting exchange with Dennett, can be found in John Searle's *Mystery of Consciousness*.

14
THE LIMITS OF COMPUTATION

Consider the following scenario. On a certain morning, the *Queen Elizabeth II* arrives at port with a gash in its hull. The damage is on the lower part of the ship and only on one side. When interviewed, the ship's crew is unsure how the damage was caused, but says that the ship took a slightly different route to get to its destination. The captain traces the ship's route on a map to demonstrate. The following day, red paint is discovered on a submerged rock along the course taken by the *Queen Elizabeth II*.

Suppose that you are investigating this case. What is the first question that you might ask? The most obvious question seems to be "What color is the *Queen Elizabeth II*?" You guessed it: the ship is red. A likely explanation for what happened, therefore, is that somewhere along its new course the *Queen Elizabeth II* came in contact with a rock that caused the gash, and, at the same time, some red paint from its hull rubbed off on the rock.

Though some might take longer than others, anyone with an ounce of common sense can come up with this explanation for what happened to the ship. Even though most people have never encountered such a scenario, the explanation is awfully simple to comprehend. What we may not realize, however, is just how much knowledge it takes to come up with this solution. We need to know, for example, that ships float. Yet, a large ship like the *Queen Elizabeth II* would not sit completely on top of the water; much of its underbelly would be submerged. That would make the ship's hull vulnerable to impact with a submerged object. We also have to understand that rocks do not float, that rocks can be sharp, that ships can move, that ships have steel sides, that steel is hard, that moving ships can collide with solid objects, that water is not solid, that rocks are hard and able to cause gashes in ships' hulls, that gashes in steel hulls can remain to be examined, and that a gash in steel plate is a possible sign of impact. That is just a fraction of the required knowledge. In addition, we need much more information about paint, sea travel, buoyancy, water, ships, steel hulls, rocks, navigation, safety precautions, speed, collision, and the behavior of ship captains and sailors. There are far too many necessary facts for me to list. Nevertheless, we discover the most likely solution to the problem and, furthermore, find it easy to figure out.

In his book *What Computers Can't Do* (later edition called *What Computers Still Can't Do*), philosopher Hubert Dreyfus says that a symbol-manipulating computer could never achieve this kind of reasoning. He argues that computers are only successful at solving problems with a limited scope. They can manipulate formulas and follow a series of steps to find the answer to a *well-defined* problem. This is the reason why computers are much better than human beings at arithmetic—arithmetic has a strict and limited set of rules. This is also why computers can beat us in chess. As complicated as the game of chess is, Deep Blue can be successful at it because the game is limited by its strict, rule-based structure. Recall the conversation program ELIZA. The programmer decided to make ELIZA a virtual therapist, as opposed to a general conversation program, because that limits the likely types of questions posed by the user. If the programmer knows that the user will ask questions about certain problems with family or friends or personal life, the programmer is better able to predict the incoming questions and prepare responses.

Even Kismet, the so-called sociable robot, runs on a set of formal instructions. This fact may not be obvious to some because Kismet responds to stimuli with facial expressions instead of written responses like ELIZA's. Nevertheless, Kismet, like every other system run by computers, follows a program. The instructions that Kismet follows might include commands to initiate its happy expression when a colorful object is detected and its fearful or surprised expressions when a fast motion is detected. Kismet follows a strict set of rules like every other computer.

A world of blocks. Dreyfus says that computer-symbol manipulation works only in a restricted field of possibilities, like this one.

Dreyfus cites a robot that responds to vocal commands to pick up and move around colored blocks and pyramids. For example, if someone says, "Pick up the red block," it will do so after responding, "OK." However, if someone says, "Grasp the pyramid," the robot will respond, "I don't understand which pyramid you mean" because there are several pyramids. Because of this seemingly intelligent behavior, AI researchers might say that the robot understands the instructions. Dreyfus disagrees because that world of blocks in which the robot operates is limited, unlike the real world in which human beings live. He says that no matter how many limited problems a machine can solve, it isn't conscious because it cannot solve a problem without boundaries, such as that of the *Queen Elizabeth II*.

Imagine that a computer and a person are both trying to predict the outcome of a horserace. This is something we would expect a computer to do well at because, like the block world, the domain of reasoning is restricted. In betting on a horse, one can often succeed only by looking at the horse's age, record, and jockey (rider) as compared to its competitors. With a good program to weight the various factors and calculate which horse has the best chance, the computer may do well, even better than the person does. However, Dreyfus says, if any other factors come into play that the computer is not specifically programmed to handle, the computer will be unable to compensate. The horse it bets on might be allergic to the nearby plants. The jockey might have just had a fight with the owner. The jockey might not do so well today if his mother died yesterday. The computer will not be able to take these factors into account. It would have to be programmed with all the subtleties of veterinary medicine, human behavior, and the nature of tragedy and sorrow, among other things. The human being, even though he may not have experienced this situation before, understands these things without preparation and can place his bet more effectively than the machine.

Again we find that a human being is able to find the solution to a problem that requires innumerable facts. The computer is programmed with formulas to predict horserace results, but will fail if a factor that is not accounted for by its program comes into play, such as the loss of the jockey's mother. For a computer, all the rules and possibilities must be well-defined. The person who is betting on the horse does not need any training to understand that the jockey will not do as well. No one has to tell him: "If the jockey's mother died recently, he will not perform as well," but that idea is obvious to him anyway. When he was first introduced to the concepts of death and loss, the person had no idea that he would one day apply them to winning a bet on a horserace (if he did, he probably would have bet a lot more money). He makes that connection on his own, based on his experience.

When world chess champion Gary Kasparov plays against Deep Blue, he calls on relevant facts from his experience. While Deep Blue is manipulating meaningless symbols based on strict rules, Kasparov is picturing the game he played a few years back when the board looked something like this. He is contemplating information from many aspects of his life that relate to chess. Irrelevant facts, such as the weight of the chess pieces, do not even enter his mind. However, if he were asked to *manufacture* chess pieces, he would know that the weight of the pieces might be relevant, even though he has never been trained in the task. He was not specifically *prepared* with the idea that weight might be a consideration when creating chess sets. He just knows what is important. His mind makes that connection. According to Dreyfus, a computer would be unable to do this.

Dreyfus, unlike Dennett, does not believe that we are machines because human beings do not, like machines, operate only by using *algorithms*—precise sets of step-by-step instructions or procedures for solving a problem. He says that those who believe we are machines assume that something that works systematically must be based on a firm set of regulations. AI researchers, for example, might say that when we make decisions based on facts that we know, we are using algorithms the same way computers do. Recall the CYC program from chapter 6. CYC's programmers have decided that the best way to duplicate human reasoning would be to create a system that formalizes human common sense as a set of rules. Dreyfus asserts that such an approach to human reasoning is a dead end. The fact that human behavior appears consistent does not mean that it is managed by formal rules. Furthermore, he says, that would be impossible. It is absurd to think that a system based only on rules and formulas would be able to solve limitless, indefinite problems such as that of the *Queen Elizabeth II*. Our grasp of ships and paint, anger and sorrow, is not stored as a list of countless facts. Human reasoning is not algorithmic.

Dreyfus says that as long as they follow formal rules, machines will only be able to solve well-defined problems, and will not achieve human reasoning. This is why, Dreyfus says, artificial intelligence researchers will never construct a conscious machine and why progress toward this goal has stalled.

He mentions one AI researcher whose program demonstrates the problem of trying to duplicate our conscious thought. The program cannot differentiate between "degrees of weirdness" when analyzing situations. For example, if the program were instructed to examine a story about a family at a restaurant, it would find it equally weird for the waiter to say the restaurant is out of food as it would be for the family to feast upon the waiter in response. Another critic of AI points to the weaknesses of computer programs used to diagnose people with illnesses based on the symptoms they experience. He says that their rule-based systems will often do well because diagnosing illness can be defined in terms of rules, but would be unable to adjust to new situations. For example, the program might diagnose a rust-spotted car with smallpox or a banana with jaundice. Machines that follow formal programs will never, according to Dreyfus, have the flexible, broad cognitive ability of the human mind.

Degrees of weirdness. We don't have to be told which scenario is more unusual, but a computer does. Computers depend on algorithms.

The well-defined architecture of computer programs also prevents them from interpreting nonliteral ideas. For instance, a computer could not figure out a metaphor like this one: "A comfortable sofa is fertile soil for the couch potato." A computer programmed to analyze writing would not pick up the nonliteral meaning of the sentence. If asked, "Is the sofa fertile soil?" a well-made program might respond "yes," but we know that this is incorrect (unless someone *really* likes nature) because it is clear to us that the statement is figurative. A more difficult question for the computer could be "Why is a sofa 'fertile soil' for a couch potato?" The computer might place couch potatoes in the same category as mashed potatoes or chopped potatoes. We could easily answer this question without preparation, but Dreyfus says that the computer could not.

Speaking of couch potatoes, let's look at a different example. A mother comes into the family room and sees her son slouching in front of the television, video game controller in hand. There are stale potato chips littering his shirt and candy wrappers scattered all over the carpet. The mother shakes her head and says, "Andrew, you look like you are working much too hard. Why don't you take a little break from all that homework and go out to play?" If this situation is analyzed by a computer and a person, only the person will come up with the correct interpretation. Computer algorithms would not be able to detect the mother's sarcasm.

Figurative language, such as metaphor, might be difficult for a child to grasp, but even a little kid can demonstrate basic understanding of a simple story. Consider this well-known fable:

> There was once a clever fox who had always used his sharp senses to find food, wherever it was. One evening, he was walking beneath an old oak tree when he noticed an enticing scent in the air. He looked above him and saw a black crow perched on a high branch with a large piece of cheese held in her beak. The fox considered this for a moment and decided that if he could trick the crow into opening her beak, she would drop the cheese and he would be able to claim it.
>
> "My, my . . . ," remarked the fox, "never have I seen such a magnificent bird. Her feathered coat is a sleek and lustrous black and her polished beak gleams in the moonlight."
>
> The crow was charmed by such praise and began to stand taller and ruffle her feathers, eagerly listening.
>
> "What noble wings this crow has," continued the fox, "and what radiant eyes. I am privileged to have seen a creature of such majesty. Surely her voice is as beautiful as her appearance. If only I could hear her sing but once, I would go home knowing that I have indeed encountered the most elegant and spectacular being I have seen in all my days."
>
> Much enamored, the shallow crow could not resist this occasion to receive further praise. She opened her beak wide and, as the cheese fell to the ground, she uttered a harsh "caw, caw!"

Smiling, the fox caught the falling cheese in his mouth and swallowed it, licking his lips contently. He looked up at the crow standing glumly on the branch and said, "You may have a voice, but sadly you do not have a brain."

Suppose that this story is given to a computer and a child to read. Each has to answer the following three questions about it:

1. Does the fox really believe that the crow is beautiful?
2. Why does the fox ask the crow to sing?
3. What is the moral of the story?

These questions are easy enough that even a child can answer them without ever having seen the story before. A computer, according to Dreyfus, would not be able to answer them unless the story and the questions were precisely programmed beforehand. It would, for example, detect all the admiring words used to describe the crow and deduce that the answer to question 1 is yes (assuming it could analyze the question or the story at all). However, even a child would be able to understand that the fox is merely flattering the crow to get the cheese. The child needs no focused training to figure this out. As for questions 2 and 3, Dreyfus says that a computer's programming would be helpless to answer them. Again we find that computer processing is ineffective for duplicating human thought because it is algorithmic whereas human thought is not.

The objection that Dreyfus presents is old. He first proposed it in 1972, eight years before the publication of Searle's Chinese Room argument. His argument caused a torrent of controversy. Researchers in the AI field criticized his objection as being short-sighted and oversimplified. They said that better machines and new programming methods would overcome the mistakes of older models. Technologists like Ray Kurzweil still say that. A stall in the progress of AI research, he says, is no reason to doubt that we will build a mind in the coming years.

There were, after all, many who thought that the Human Genome Project would take centuries to complete, when it actually took only thirteen years, finishing ahead of schedule. There were many others who doubted that cloning was possible, but they were silenced when the sheep Dolly was cloned successfully. Our perception of human life was transformed by the discovery of DNA, despite whatever critics of Watson and Crick might have said. Almost every belief about what will occur in the future or about something that cannot yet be proven will have its share of critics. Kurzweil stands by his "theory of accelerating returns," holding that the accelerating growth of technology will make possible the emergence of superintelligent machines. Kurzweil asserts that we are on the verge of a new age of technological growth. New methods and innovations will allow us to accomplish things that we cannot even imagine today. If the current approach to machine consciousness fails, we will find another one that succeeds.

Dennett still unwaveringly advocates that there is no doubt that we will construct machines with as many, or more, abilities than we possess because we have been machines all along. He dismisses the arguments of Searle and Dreyfus because he denies that consciousness exists at all.

Then there are the scientists who believe that we are machines—biological machines—because they say that our organic machinery controls what we do. Francis Crick, for example, uses evidence from alien hand syndrome and other brain disorders to show that what we call "free will" is merely the activity of certain brain regions. The brain is programmed to calculate our decisions. How can human reasoning not be algorithmic? Neurons work in binary fashion, just as circuits do. We are organic machines, and if an organic machine can be intelligent, so can a man-made machine.

Despite such claims, Dreyfus's objection has not yet been overcome. Then again, it has not been proven. We don't know what new technologies or scientific discoveries will emerge in the coming years. Dreyfus is certain that time will make no difference because there is simply no way that computer calculations can generate human nonalgorithmic reasoning. We can think in ways that computers cannot. We can understand and theorize about unlimited problems such as that of the jockey and the *Queen Elizabeth II*. Even as circumstances change, we can decide what aspects of our vast experience are relevant to the situation at hand. We are able to discover things that are not precisely spelled out for us. These abilities, Dreyfus says, would be impossible through algorithm alone. Human reasoning works in a very different way; we need only to find out how.

FURTHER READINGS

The scenario of the *Queen Elizabeth II* that begins this chapter was taken from a paper by Larry Wright titled "Argument and Deliberation: A Plea for Understanding," published in the *Journal of Philosophy*. To learn more about the work of Hubert Dreyfus, look for his book *What Computers Still Can't Do*. The ideas of the jockey, the block-manipulating robot, and degrees of weirdness were taken from there. An interesting discussion of Dreyfus's objection, Searle's Chinese Room, and artificial intelligence can be found in the *Scientific American* article "Could a Machine Think?" by Paul and Patricia Churchland.

15

A NEW GENERATION

Imagine that a hot-air balloonist named Rocco embarks on a journey from Nova Scotia to meet a friend of his in France. Over the next few days there is a violent storm over the Atlantic Ocean. Pouring rain and thundering waves divert nearby vessels. On the day of the meeting, Rocco does not arrive at his friend's house and the coast guard mobilizes for a search. After several days of searching, Rocco's balloon is found, torn and flooded, five hundred miles from the European shoreline.

We know three things about Rocco that suggest what might have happened to him:

1. Rocco set out in a balloon from Nova Scotia traveling over the Atlantic.
2. A violent storm took place over the Atlantic.

3. Rocco's empty balloon was found flooded and five hundred miles from his destination.

What we have here is an unrestricted scenario much like that of the *Queen Elizabeth II* in the previous chapter. In the previous case, we decided on the explanation that seemed most obvious. This time, however, we will use what we know to decide upon *several* possible explanations of what happened to Rocco. Here are some possibilities, of varying likelihood, that we might come up with:

A. Rocco drowned in the ocean.
B. He was rescued by a passing freighter.
C. He swam to shore.
D. He is still treading water.
E. He was kidnapped by pirates.
F. He was teleported to another galaxy.
G. He is living inside a whale, coexisting with it in a mutually beneficial, symbiotic relationship.

The most plausible explanation here, unfortunately for Rocco and his friend, is A. Explanation B is the next most likely. One could argue whether F or G is least likely. All the explanations are possible, though, and we can imagine them without using algorithms or being specifically trained for the task in advance. We also can easily decide which explanations are most and least likely without knowing any numerical probabilities.

We already saw that, according to Dreyfus, computers lack the ability to come up with answers to such unlimited problems. Since we do have that ability, our reasoning is different from that of a computer; it is not algorithmic. Now that we know what human reasoning is not, we need to determine what it is, and why algorithms cannot duplicate it.

It is here that I will make *my* contribution to the controversy by expanding upon the theories of some of the individuals we have discussed, and rejecting those of others, to suggest an idea of human reasoning that answers the question posed by the title of this book. Before we address what human reasoning is, however, let's for a moment return to John Searle's Chinese Room.

Searle concludes that symbol-manipulation (algorithmic reasoning) is insufficient for understanding because one could imagine a man in a room correctly answering questions in Chinese by manipulating symbols according to a rulebook without actually being able to understand any Chinese at all. Recall that his logic works as follows:

1. Computer programs work by manipulating symbols.
2. Minds have mental contents (consciousness) and allow for understanding.
3. Symbol manipulation is insufficient for understanding and consciousness.
4. Therefore, executing a computer program is insufficient for a mind.

I agree with this assessment, but there is something very important that Searle leaves out. A major criticism of Searle's argument is that step 3 of his argument is an assumption that lacks proof. How does he know that symbol manipulation is insufficient for understanding? Ray Kurzweil, Marvin Minsky, and other proponents of AI research certainly do not agree that such a claim can be assumed. That is what AI is trying to accomplish. Searle does not explain *why* manipulating symbols cannot generate understanding.

I suggest that the reason why symbol manipulation (algorithm) is insufficient for consciousness is that, as Dreyfus says, manipulating symbols can only find solutions to well-defined, limited problems, whereas human consciousness can be used to solve solutions to ones that are boundless and unfettered. Because the algorithmic reasoning of a computer program cannot do everything that the nonalgorithmic human mind can, algorithmic reasoning alone is not enough for consciousness to exist. Here is how I formally explain step 3 of Searle's argument:

1. Symbol-manipulation is rule-bound.
2. Consciousness is boundless.
3. A rule-bound system is not sufficient for a boundless system.
4. Therefore, symbol manipulation is not sufficient for consciousness.

Above, I use Dreyfus's idea to explain what Searle does not: the third step in his Chinese Room argument. As I understand it, this combination has not been suggested by others. Searle and Dreyfus have always been viewed as having two separate categories of objections to Strong AI (the idea that a properly programmed computer can generate consciousness). Searle's is often called the "internal" objection because it maintains that though the computer may seem conscious on the outside, it will lack understanding and mental contents *on the inside*. On the other hand, Dreyfus's is called the "external" objection because it says that, *on the outside*, a computer could never display the creative and flexible abilities of the human mind, such as discovering the explanation of the *Queen Elizabeth II* scenario, the jockey scenario, or Rocco's scenario. Perhaps the two objections should not be placed in separate categories, but combined. This way, we show that algorithmic symbol-manipulation is not enough to generate consciousness because it cannot do many things that the mind can do.

We are back to where we were before: we know what human reasoning is not. The next step is to explain what it is and what makes it so different from the reasoning of computer algorithms.

Consider the way we explain Rocco's scenario. We decide that the most likely occurrence is that Rocco drowned in the ocean. Such a problem cannot be defined in terms of rules, yet we can all come to this same conclusion. How do we figure it out? Behind the mysterious power of consciousness is enormous, immeasurable background knowledge. Each of us has a detailed model of how the world works. From our knowledge of the world, we gain insight into things that happen in the world and are able to reason.

One might say that the set of rules used to program a computer serves just as well as a representation of the world. I disagree. Let me suggest that our mental model of the world is not stored in a set of formal rules, but rather in what I will call "binding qualia."

One aspect of the consciousness debate that we have not met before in this book is what's called the "binding problem." Philosophers and neuroscientists wonder how, when we perceive the world, the different types of qualia bind together to form a single cohesive experience. For example, during a day at the beach, the feeling of the sand between your toes, the warm sunlight on your back, and the sound of rolling waves all bind together, even though they are associated with different receptors, different neural pathways, and different parts of the brain.

Just as individual senses combine to form the qualia of looking at the ocean, qualia from all parts of our lives (binding qualia) combine to form our mental model of the world from which human reasoning is derived. An example of simple qualia is *what it is like* to hear waves rolling in. An example of binding qualia would be what it is like to spend a day at the beach. The binding qualia of a day at the beach might bind to that of the person you were at the beach with. The qualia of both those experiences might bind to that of your cousin's wedding, which you were thinking about while at the beach, as well as the qualia of intense pain in your foot because you stepped on a jellyfish while wading in the ocean. Binding qualia, in short, are larger concepts derived from simpler qualia that bind to one another to form a unified representation of the world.

Binding qualia merge based on the circumstances in which we experience them, the order of events, the connections we, with our free will, make between them, and many other factors. We associate ideas based on the way they originally bind and the way these bindings evolve throughout our lifetime. Maybe that is why we often go on thought tangents where we associate one idea with a related idea and then associate *that* idea with a different idea and so on. That might also be why there are songs that remind us of certain people or places or why we picture things in our minds when we read a novel. That may be why some people cry when they hear the national anthem or can't help smiling when they hear the name of their old class clown.

Binding qualia form our vast mental model of the world from which we can draw relevant information and ignore irrelevant information in order to solve unlimited problems such as explaining what happened to poor Rocco. From this model, we understand that it is dangerous for Rocco to be in a balloon during a storm for two days. We understand that if his balloon is found empty and flooded in the middle of the Atlantic, Rocco is almost certainly lying in its watery depths. We know from our model that freighters travel in the ocean, but it is unlikely that they would find Rocco, especially after a serious storm. It is also clear to us, from our picture of the world, that Rocco has probably not been teleported out of the galaxy or been living in a whale. We know all this because our binding qualia coalesce into a massive, dynamic information store based on what we experience.

There is no way that algorithms can duplicate the action of binding qualia in forming our mental model of the world. Algorithms must be based on existing rules; they require *context*. An algorithm wouldn't know how to form our model of the world because it would have no basis for creating it. How would an algorithm know that sugar is supposed to taste sweet or that pain is supposed to hurt? How would it know that storms are dangerous for hot-air balloonists? In human beings, the qualia of these things bind together and give us that understanding. We don't even have a sense of logic without our experiences, and algorithms can't create a system of logic. Algorithms follow rules; they cannot form them.

That is why I don't think the CYC program has or will ever have human reasoning. The CYC system, remember, is a program with over one million rules that tries to simulate human reasoning. The project began in 1984, before I was born, as a team of programmers added rule after rule. After all this time, CYC is not even close to having human common sense. I didn't have a team of people teaching me about the world every day. I gained my understanding of the world through binding qualia.

Even if we have never flown in a hot-air balloon, we know that a storm would be a danger because we can draw from our vast mental model. We are able to think of situations that we have not yet experienced by using our mental model as a foundation. This we call *imagination*.

Our imagination can feel real at times because, just as there is something it is like to be a bat or to see red, there is something it is like to be *thinking*. Furthermore, there is something it is like to be thinking about thinking. There are qualia associated with our use of the mental model. I believe that this is what gives us our ability to be creative. This is how we can be confused and conflicted, humorous and philosophical. It is how we can come to realizations, draw conclusions, and make decisions. The set of qualia associated with human thinking is what consciousness is.

So, not only do binding qualia provide us with comprehensive background knowledge, they are also an active part of our everyday thought processes. Binding qualia are the key to human reasoning. That is why, I believe, a programmed computer cannot have consciousness.

Let's now see how my view compares to the views of all those we have discussed. My theory contrasts with that of Daniel Dennett, who says that we *are* algorithmic machines and that consciousness and qualia do not exist. I also do not agree with Kurzweil and Minsky when they say that we can program a conscious machine. Algorithm is insufficient for consciousness.

Unlike David Chalmers, I do not think that zombies are logically possible because, without consciousness, they cannot demonstrate human reasoning and solve problems like that of Rocco and the *Queen Elizabeth II*. There could not be zombie philosophers or detectives (such as Sherlock Holmes). Algorithmic systems would fail in such professions.

I do not accept the Turing Test as a valid means of detecting consciousness for the same reasons that Searle presents. Finally, I like Frank Jackson's thought experiment of Mary in the black and white room. My theory strongly involves qualia, and Jackson's thought experiment stresses their importance. In my view, the role of qualia in our reasoning is what makes us human.

So, if a scientist mastered the workings of your brain, I don't think that he would know everything about your mind. He would not understand what it is like to have your consciousness. He would not be able to predict your actions. Unlike Francis Crick, I believe in free will. If our physical structure were to determine our thoughts and actions, we would be subject to its rules—we would be dependent on algorithms. I do not believe that anyone can have access to the consciousness of another because qualia are private, inner experiences.

Similarly, I disagree that computer technology can be used to build a mind. Other methods may be used to create conscious machines, as long as the result is an entity not limited by formal rules. If a machine is constructed so that it has human reasoning, it will no longer be a machine, but I highly doubt that this will ever happen. With all of this in mind, my answer to the question posed by the title of this book is no. In my view, we all *depend* on machinery, but we are not machines.

Of course, just as we discussed the opinions of scientists and philosophers, all of this is my opinion. It is interesting to note how the views of consciousness throughout this book are so different. Chalmers, Ryle, Crick, Edelman, Kurzweil, Turing, Minsky, Searle, Jackson, Damasio, Dennett, and Dreyfus are all very intelligent and knowledgeable people. Yet, none of them can agree on a single view. It is even possible (and likely) that all of them are wrong, at least in part. The debate is far from over.

The controversy over the mind-body problem dates back to the philosopher Descartes in the seventeenth century, making the question of consciousness nearly four hundred years old. However, in hundreds of years of discussion, philosophers have only begun to advance beyond the simple ideas of dualism and materialism.

In the last fifty years, the controversy has come a long way. Because of modern advances in computing and neuroscience, we can now view the issue from revolutionary new angles. As we approach the middle of the twenty-first century, we will see whether the predictions of Kurzweil, Dreyfus, and others hold true. We will get closer and closer to the answers.

Though the scientists and philosophers theorizing about consciousness are far from finding the answer, their difficult debate provides us with a better understanding of ourselves. As these theories clash and evolve, we move closer to unraveling the mystery by creating openings for new ideas. We are on the verge of a new generation of theories and theorists.

Though I have answered the question of whether we are machines, I have not proposed a theory to explain the existence of qualia or the nature of consciousness. That problem remains. The fact that so much about consciousness is still unknown has sparked a new interest in the subject. More and more research is being devoted to studying consciousness, as it is likely the greatest mystery we now face. More than once, throughout history, a single discovery has changed the world. The accidental discovery of penicillin revolutionized medicine. The invention of the Internet restructured the way we live. The discovery of DNA transformed the study of biology. Momentous advancements as those were, they will seem almost trivial when compared to a set-tled account of how the human mind works. If it is our genera-tion that solves the mystery of consciousness, stirring events are on the horizon.

FURTHER READINGS

The scenario of Rocco and his hot-air balloon, including most of the possible solutions that I listed, is from the book *Better Reasoning: Techniques for Handling Argument, Evidence and Abstraction* by Larry Wright. For further information about why human thought is not algorithmic, try *What Computers Still Can't Do* by Hubert Dreyfus. There are also hundreds of other great books on consciousness out there, if you are inter-ested in going further.

BIBLIOGRAPHY

"ALICE Finalist in 2004 Loebner Contest." Alicebot.org. Available at http://www.alicebot.org/ (accessed September 10, 2004).

Angel, Leonard. *How to Build a Conscious Machine*. Boulder, CO: West-view Press, 1989.

Atwood, Margaret. "My Life as a Bat." In Atwood, *Good Bones and Simple Murders*. New York: Nan A. Talese, 2001, pp. 109–16.

Baars, Bernard J. *In the Theater of Consciousness: The Workspace of the Mind*. New York: Oxford University Press, 1997.

Behar, Michael. "The Doctor Will See Your Prototype Now." *Wired*, February 2005.

Blackmore, Susan. *Consciousness: An Introduction*. Oxford, UK: Oxford University Press, 2004.

Block, Ned. "Troubles with Functionalism." *Minnesota Studies in the Philosophy of Science* 9 (1978): 261–325.

Bostrom, Nick. "When Machines Outsmart Humans." NickBostrom.com. Available at http://www.nickbostrom.com/2050/outsmart.html (accessed January 4, 2005).

Chalmers, David J. *The Conscious Mind: In Search of a Fundamental Theory.* Oxford, UK: Oxford University Press, 1996.

———. "Consciousness and Its Place in Nature." In Chalmers, *Philosophy of Mind*, pp. 247–72.

———. "Facing Up to the Problem of Consciousness." *Journal of Consciousness Studies* 2, no. 3 (1995): 200–19.

———. *Philosophy of Mind: Classical and Contemporary Readings.* Oxford, UK: Oxford University Press, 2002.

"Chess." Brainencyclopedia.com. Available at http://www.brainy encyclopedia.com/encyclopedia/c/ch/chess_1.html (accessed April 25, 2004).

Churchland, Paul M., and Patricia S. Churchland. "Could a Machine Think?" *Scientific American*, January 1990.

Crick, Francis. *The Astonishing Hypothesis: The Scientific Search for the Soul.* New York: Charles Scribner's Sons, 1994.

Crick, Francis, and Christof Koch. "Consciousness and Neuroscience." *Cerebral Cortex* 8 (1998): 97–107.

Daly, James. "Soul of a New Machine." Business2.com. Available at http://www.business2.com/b2/web/articles/0,17863,527113,00 .html/ (accessed January 4, 2005).

Damasio, Antonio R. *Descartes' Error: Emotion, Reason and the Human Brain.* New York: Avon Books, 1994.

"Deep Blue." IBM.com. Available at http://www.research.ibm.com/ deepblue/meet/html/d.3.html (accessed March 21, 2004).

Dennett, Daniel C. *Consciousness Explained.* Boston: Little Brown and Company, 1991.

———. "Consciousness in Human and Robot Minds." In Ito, Miyashita, and Rolls, *Cognition, Computation and Consciousness.* Also available online at KurzweilAI.net.

———. *Freedom Evolves.* New York: Viking Press, 2003.

———. "Quining Qualia." In Chalmers, *Philosophy of Mind*, pp. 226–46.

Dreyfus, Hubert L. *What Computers Still Can't Do: A Critique of Artificial Reason.* Cambridge, MA: MIT Press, 1992.

Eccles, John C. *Mind and Brain.* New York: Paragon House Publishers, 1985.

Eccles, Lisa. "MIT Scientists Create a More 'Sociable Robot.'" Elecdesign.com. April 30, 2001. Available at http://www.elecdesign.com/Articles/ArticleID/4005/4005.html (accessed April 4, 2004).

Edelman, Gerald M. *Bright Air, Brilliant Fire: On the Matter of the Mind.* New York: Basic Books, 1992.

———. *The Remembered Present: A Biological Theory of Consciousness.* New York: Basic Books, 1989.

———. *A Universe of Consciousness: How Matter Becomes Imagination.* New York: Basic Books, 2000.

Guyton, Arthur C. *Anatomy and Physiology.* Philadelphia: Saunders College Publishing, 1985.

Hanley, Richard. *The Metaphysics of Star Trek.* New York: Basic Books, 1997.

Heil, John. *Philosophy of Mind: A Guide and Anthology.* Oxford, UK: Oxford University Press, 2004.

Hobson, J. Allan. *Consciousness.* New York: Scientific American Library, 1999.

Ito, Masao, Yasushi Miyashita, and Edmund T. Rolls. *Cognition, Computation and Consciousness.* Oxford, UK: Oxford University Press, 1997.

Jackson, Frank. "Epiphenomenal Qualia." *Philosophical Quarterly* 32 (1982): 127–36.

Kelly, Clinton W. "Can a Machine Think?" KurzweilAI.net. Available at http://www.kurzweilai.net/meme/frame.html?main=/articles/art02 14.html?m%3D4 (accessed January 10, 2005).

Kurzweil, Ray. *The Age of Intelligent Machines.* Cambridge, MA: MIT Press, 1990.

———. *The Age of Spiritual Machines: When Computers Exceed Human Intelligence.* New York: Penguin Books, 2000.

———. *Are We Spiritual Machines? Ray Kurzweil vs. the Critics of Strong AI.* Seattle, WA: Discovery Institute, 2002.

———. "The Coming Merging of Mind and Machine." *Scientific American*, September 1999.

———. "Live Forever—Uploading the Human Brain . . . Closer than You Think." PsychologyToday.com. Available at http://cms.psychologytoday.com/articles/pto-20000101-000037.html (accessed January 8, 2005).

———. "My Question for the Edge: Who Am I? What Am I?" Edge.org.

Available at http://www.edge.org/q2002/q_kurzweil.html (accessed January 3, 2005).

LeDoux, Joseph. *Synaptic Self: How Our Brains Become Who We Are*. New York: Penguin Books, 2002.

Lenat, Douglas B. "CYC: A Large-Scale Investment in Knowledge Infrastructure." *Communications of the ACM* 38, no. 11 (1995): 33–38.

Lenat, Douglas B., Ramanathan V. Guha, Karen Pittman, Dexter Pratt, and Mary Shepherd. "CYC: Toward Programs with Common Sense." *Communications of the ACM* 33, no. 8 (1990): 30–49.

Lewis, David. "What Experience Teaches." In *Proceedings of the Russellian Society*. Sydney, Australia: University of Sydney, 1988.

McCarthy, John. "What Is Artificial Intelligence?" Computer Science Department, Stanford University, 2000.

McGinn, Colin. *The Mysterious Flame: Conscious Minds in a Material World*. New York: Basic Books, 1999.

McNally, Phil, and Sohail Inayatullah. "The Rights of Robots: Technology, Culture and Law in the 21st Century. Metafuture.org. Available at http://www.metafuture.org/Articles/TheRightsofRobots.htm (accessed January 17, 2005).

Minsky, Marvin L. "Conscious Machines." *Machinery of Consciousness*. Proceedings of the National Research Council of Canada, 75th Anniversary Symposium on Science in Society, 1991.

———. *The Society of Mind*. New York: Simon and Schuster, 1985.

———. "Will Robots Inherit the Earth?" *Scientific American*, October 1994.

Moody, Todd C. "Conversations with Zombies." *Journal of Consciousness Studies* 1, no. 2 (1994): 196–200. http://www.imprint.co.uk/Moody_zombies.html.

"Moore's Law." Intel.com. Available at http://www.intel.com/research/silicon/mooreslaw.htm (accessed April 18, 2004).

Moravec, Hans. "Letter from Hans Moravec." *New York Review of Books*, March 25, 1999. Also available online at KurzweilAI.net, http://www.kurzweilai.net/meme/frame.html?main=/articles/art0017.html? (accessed November 29, 2006).

Nagel, Thomas. "What Is It Like to Be a Bat?" *Philosophical Review* 83 (1974): 435–50.

Popper, Karl R., and John C. Eccles. *The Self and Its Brain.* Berlin: Springer-Verlag, 1977.

Preston, John, and Mark Bishop. *Views into the Chinese Room.* Oxford, UK: Oxford University Press, 2002.

Rychlak, Joseph F. *In Defense of Human Consciousness.* Washington, DC: American Psychological Association, 1997.

Ryle, Gilbert. *The Concept of Mind.* London: Hutchinson & Company, 1949.

Schank, Roger. "Can Computers Decide?" KurzweilAI.net. Available at http://www.kurzweilai.net/meme/frame.html?main=/articles/art02 21.html?m%3D3 (accessed January 10, 2005).

Searle, John R. "Consciousness." KurzweilAI.net. Available at http://www.kurzweilai.net/meme/frame.html?main=/articles/art0282 .html?m%3D3 (accessed January 19, 2005).

———. "I Married a Computer." In Kurzweil, *Are We Spiritual Machines?* pp. 56–77.

———. "Is the Brain's Mind a Computer Program?" *Scientific American,* October 1990.

———. *Mind, Language, and Society: Philosophy in the Real World.* New York: Basic Books, 1998.

———. "Minds, Brains and Programs." *Behavioral and Brain Sciences* 3, no. 3 (1980): 417–24.

———. *Minds, Brains and Science.* Cambridge, MA: Harvard University Press, 1983.

———. *The Mystery of Consciousness.* New York: New York Review of Books, 1997.

———. *The Rediscovery of the Mind.* Cambridge, MA: Bradford Books/MIT Press 1992.

Sylwester, Robert. *How to Explain a Brain: An Educator's Handbook of Brain Terms and Cognitive Processes.* Thousand Oaks, CA: Corwin Press, 2005.

Turing, Alan. "Computing Machinery and Intelligence." *Mind* 59, no. 236 (1950): 433–60.

Underwood, Geoffrey. *The Oxford Guide to the Mind.* New York: Oxford University Press, 2001.

Weizenbaum, Joseph. "ELIZA—A Computer Program for the Study of Natural Language Communication between Man and Machine." *Communications of the ACM* 9, no. 1 (1966): 36–45.

Wright, Larry. "Argument and Deliberation: A Plea for Understanding." *Journal of Philosophy* 92, no. 11 (1995): 565–85.

———. *Better Reasoning: Techniques for Handling Argument, Evidence and Abstraction.* New York: Holt, Rinehart and Winston, 1982.

Zeman, Adam. *Consciousness: A User's Guide.* New Haven, CT: Yale University Press, 2002.

INDEX